Audio Post-production in Video and Film

Second edition

Tim Amyes

Focal Press

OXFORD AUCKLAND BOSTON JOHANNESBURG MELBOURNE NEW DELHI

Focal Press
An imprint of Butterworth-Heinemann
Linacre House, Jordan Hill, Oxford OX2 8DP
225 Wildwood Avenue, Woburn, MA 01801-2041
A division of Reed Educational and Professional Publishing Ltd

Ⓡ A member of the Reed Elsevier plc group

First published as the *Technique of Audio Post-production in Video and Film* 1990
Paperback edition 1993
Reprinted 1994, 1995 (twice), 1997, 1998
Second edition 1998
Reprinted 2000, 2001

British Library Cataloguing in Publication Data
A catalogue record for this book is available from the British Library

Library of Congress Cataloguing in Publication Data
A catalogue record for this book is available from the Library of Congress

ISBN 0 240 51542 0 ✓

Typeset by Avocet Typeset, Brill, Aylesbury, Bucks
Printed and bound in Great Britain by Biddles Ltd, *www.biddles.co.uk*

Contents

Preface

The computer and digital audio recording have radically changed the audio post-production industry. The standard techniques of 50 years no longer apply. New computer-based systems, at moderate cost, can produce results that a few years ago only high-end facilities houses could hope to achieve. This book will help those with a basic knowledge of sound to understand the processes and techniques needed to audio post-produce programmes and projects successfully.

This book is intended for those looking for an understanding of post-production from location recording through to transmission or release.

Acknowledgements

In writing this book many friends and colleagues in the industry have contributed. Among these are Andy Boyde, Ian Finlayson, John Hughes (AMS Neve), Cy Jack (Waterfront Studios), Terry Kain, Gillian Munro, Louis Krammer, Mike Varley (Northern Film Labs), Alistair Biggar, Tim Mitchell, Len Southam and many others. I am also indebted to Chris Davis for his technical advice on computer technology, particularly for the appendix, and for his advice as an informed and interested layman.

The following companies have kindly provided drawings and diagrams: AMS Neve, Carlin Music, CMX, De Wolfe, Dolby Labs, Eastman Kodak, Klark-TechniK Research Ltd, JVC Professional, Sony Broadcast & Communications, Sound Ideas, Soundscape Digital Technology, Yamaha Ltd, Waterfront Studios, Glasgow.

Dolby and the double D symbol are trademarks of Dolby Laboratories Licensing Corporation.

An historical background

Since man first reproduced images on a screen he has wished to add sound to increase the illusion of reality. Thomas Edison's laboratory is credited with producing the world's first moving talking picture when Edison's assistant walked towards the Edison sound camera, lifted his hat and said, 'Good morning, Mr Edison.'

The pictures were viewed in a 'What the butler saw' device and the sound was heard through acoustic headphones. This was very much a personal viewing and listening system. There was no electrical amplification system and therefore no way of acoustically projecting the sound to an audience, and there was certainly no system to alter or improve the sound once it had been recorded. The techniques of audio post-production would not be perfected for another 30 years or more. However, in less than 10 years the basic technology needed to produce sound/motion pictures successfully had been developed.

In 1904 Eugene Lauste, a Frenchman working in London, managed to record sound onto a piece of photographic film. Now it was possible to record sound on the same piece of film as the picture, so that the two were in perfect synchronization, and to copy the soundtrack and the picture simply by printing it photographically onto another piece of film. Despite this technical breakthrough, there was still no way to amplify sound successfully, although attempts were made to use compressed air systems, with some success. It was not until Lee de Forest invented the electronic amplifying valve or tube in 1914 that successful projection of both sound and picture to a large audience was possible.

In 1926 the world's first commercially successful sound feature film, 'The Jazz Singer', was made in America, using cumbersome gramophone discs synchronized with pictures. The object was not to make a dialogue film, but to produce a film that would no longer require a live orchestra as a musical accompaniment – thereby producing a massive cost saving for the exhibitor. At the same time as these sound feature films were being made, Fox's 'Movietone News' was introduced, recording actual events as they were happening. Originally, each sound newsreel item was introduced with silent titles, but it was soon realized that voice overs and music could enliven the 'reels'. Therefore, methods were developed which would allow a narration to be added and mixed with music. In 1930 the standard Moviola editing machine was modified to allow it to run both picture and sound in synchronization, and it became possible to cut various soundtracks in synchronization. The techniques of audio post-production were being developed.

Within the next 15 years the film industry developed all the basic operational techniques needed to produce a polished quality soundtrack: techniques that are used today both in film and video productions; techniques that allow various separate sounds to be locked together in perfect synchronization, and then to be mixed together on a separate track to produce a final completed soundtrack. At first this proved difficult as background noise was increased dramatically by mixing and copying sound onto the final film. Early Laurel and Hardy films demonstrate this in their simple mixes of music, dialogue and effects. In 1931 Western Electric introduced the first photographic noise reduction process but the problem of background noise was not eliminated until the introduction of digital recording in the 1980s. Western Electric's 1929 system is shown in Figure 1.1.

In 1939 stereo sound arrived in the cinema with the Walt Disney film 'Fantasia'. Even today's multitrack music recording techniques were anticipated by Hollywood's large studios who used multiple track photographic sound recording systems in the late 1930s.

The post-synchronization or replacement of dialogue with dialogue in another language was an early development. The silent film's international market was lost with the introduction of the 'talkie' and America was particularly badly hit. The American product needed a larger market than just the English-speaking world, so post-synchronization was developed, allowing dialogue to be re-recorded in a foreign language after a film had been edited and completed. The technique was perfected in Paramount's Joinville studios near Paris.

During the Second World War there were further developments in electronics that would affect the film industry; it became apparent that Germany had developed a new and revolutionary recording system, superior to acetate disc or optical film – a system that explained how Hitler had managed to deliver speeches that were apparently live in places so far apart that he could not have had time to travel between them. The secret was magnetic recording. It was of a higher quality than even the best 'push-pull' photographic soundtracks available at the time. It was now no longer necessary to wait for photographic processing before playing back film recordings; sound could now be heard immediately. However, every move forward has a tendency to mark a step back.

Sound editors complained that, with the new-fangled magnetic sound, it was no longer possible to see the actual optical sound modulation on the track. This tended to make sound editing more difficult. (Today, with audio post-production work stations using digital sound, this facility is once again available.)

The Second World War also saw the development of computers to help crack Nazi secret codes. They were crude at first, but developed into the personal computer which eventually formed the basis of the first picture and audio workstations of the 1980s.

In the early 1950s the film industry began to feel the competition of televi-

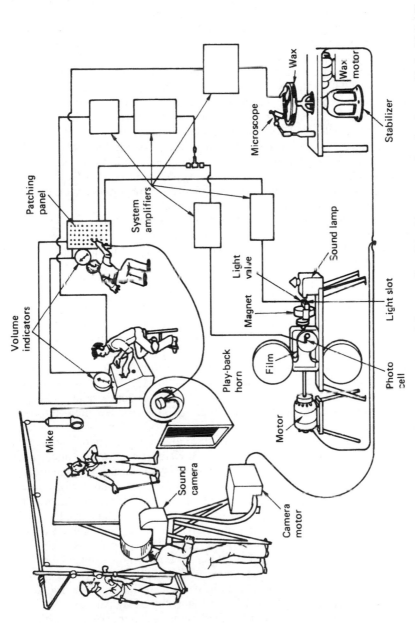

Figure 1.1 The Western Electric sound recording system of 1929 film and disc recording. 'Discs' have returned to the cinema with the Digital Theatre System's separate CD-ROM disc projection system

sion and it responded by producing wide-screen pictures, with stereophonic sound recorded on magnetic sound tracks bonded to the picture film.

These early stereo feature films were produced in true stereo. The dialogue was recorded in stereo on the set. This meant that each time the picture cut, the dialogue moved its screen position – perhaps even in mid-sentence – as one picture was edited onto the next. This discontinuous stereo sound was unacceptable. It became standard practice to record dialogue in mono on location and to move it about the large sound field the cinema provides, in the audio post-production process.

Early stereo magnetic striped films had their problems too, for they tended to shed oxide and clog the projector's sound heads, muffling the sound (magnetic stereo release prints are no longer used) and so cinema owners returned to replaying poor-quality monophonic optical soundtracks, with the extra amplifiers and loudspeakers lying redundant and unused. In 1971 stereo sound once more became available but this time pursuing the more practical photographic recording techniques. The Dolby Corporation offered a photographic recording system providing four separate soundtracks – Dolby stereo. The system was ideally suited to cinemas which were already wired for the redundant magnetic four-channel sound systems.

In the home, the quality of sound reproduction was improving. The long-playing record and the hi-fi system had arrived but, perhaps because of lack of interest from the consumer, the television set still sat in the corner of the room producing lo-fi sound. Indeed, manufacturers had tried hard even in the 1930s to increase television sales by offering quality wireless receivers capable not only of long and medium wave reception but also picking up VHF television sound. These efforts were unsuccessful, as were the attempts to offer full sound radio facilities on television receivers. By the late 1940s, television sound transmission had been relegated to being the Cinderella of broadcasting. The total audio amplification part of a receiver was often just one valve, the pictures were small and the sound limited. Styling restrictions and the effects of stray magnetic fields from the loudspeaker onto the picture tube all resulted in the use of small speakers and poor sound quality. This produced a situation which did not encourage technical or aesthetic improvements. Figure 1.2 shows the operation of the Cinerama film.

Television was a live medium. It was quite unlike film, where events were recorded and painstakingly re-created; a television programme, once transmitted, was lost forever. In television there was no minute examination of detail and no post-production situation. A repeat was a second performance – a complete re-creation of a live programme. Like a theatrical production, the TV show either had to go on or just go off the air – a situation completely different from film making. In 1955 the situation began to change; the BBC went on the air in London with a recorded programme. The recording machine was called Vera (Video Electronic Recording Apparatus) and used a high tape speed of 200 inches/second (as opposed to the standard audio speed of 15 inches/second) but the quality was poor. Two years later the world's first quality commercial videotape recorder was introduced by Ampex.

The Ampex videotape recorder produced pictures capable of a definition up to four million cycles per second. It was originally intended to allow live television programmes to be recorded on America's east coast, and then time-delayed for transmission at the 'same time' hours later, on the west coast. This

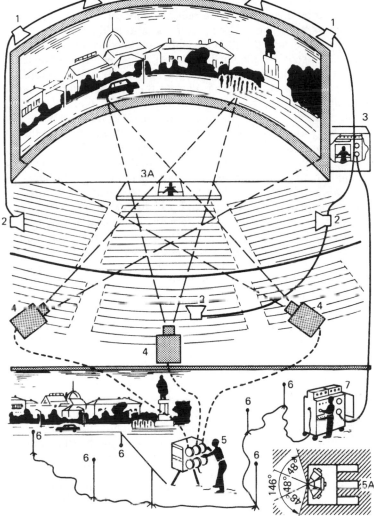

Figure 1.2 The Cinerama film: 1, left, mid, centre rear loudspeakers; 2, ambient surround speakers; 3, projection sound control; 4, three synchronized projectors; 5, camera; 5a, camera arrangement; 6, microphones for stereo sound; 7, sound mixing

system spawned two new techniques in television editing and audio post-production. Originally, a razor blade was used to cut and edit videotape, and the ground-breaking 'Rowan and Martin's Laugh In' was made this way in the 1960s, but this method was soon replaced by systems designed for editing by copying the pictures from one machine to another.

With the introduction of video picture editing it became apparent, as the film industry had discovered in 1927, that the soundtrack also needed editing and 'sweetening'. Since it was very expensive to tie-in film sound editing systems to videotape, another system was required – one that would be as flexible and versatile as film. The multitrack tape recorder from the music studio provided an answer. It could record many different soundtracks all in synchronization and linked to the picture, and to achieve this a special electronic code was developed, rather like the sprockets used in film – time code.

The shrinking size of videotape equipment is shown in Figure 1.3.

In the mid-1980s digital audio recording became available to music studios and through digital audio compact discs to the general public at home. This is the ultimate in recording quality, a quality which is available with pictures and surround sound on the digital versatile disc (DVD) and closely duplicated in the domestic video cassette recorder offered with a hi-fi capability. It is for the domestic environment that the vast majority of soundtracks are created in audio post-production, from television surround sound to computer games – and to complement the digital quality sound there are digital quality pictures. However, the word digital applied to audio and video needs to be clarified before the term quality can be applied, for if digital compression is applied to a signal, quality may be at risk. This happens in most domestic systems.

Audio post-production systems offer digital recording most often within a digital workstation, often using digital audio consoles. This is a particularly appropriate system since almost perfect sound quality is matched by almost instant access. No longer is it necessary to spool through many feet of tape or sound film to locate specific items – access is instant.

The computer has changed many aspects of our lives and audio post-production is no exception. It is no longer always essential to hire expensive audio suites to 'sweeten' soundtracks. Just as musicians now have their own computer workstations to compose and record their compositions, so editors and mixers can create and mix their soundtracks in a controlled home studio environment. The difference between video and film as an origination format is much less than it used to be. Film can be given a video look and digital video can be given a film look, arguably indistinguishable from the 'real thing'! Certainly for cinema projection film has many advantages over high-definition video systems. But the vast majority of productions are shown in the home. Unsurpassed sound quality in digital formats is now available from terrestrial and satellite transmitters to the home.

One of the first digital audio workstations is shown in Figure 1.4.

From a production point of view the ability to transmit high-quality pictures and sound around the globe with ease means that material no longer has to be physically shipped from one location to another. A production can be edited in one country while the special video effects are produced in another and the music is prepared and recorded in a third. Through modern technology it is possible for producers to view, listen and communicate with technicians about their work without ever meeting them. The cost savings can be enormous.

The technical aspects of audio post-production have changed dramatically over the years. However, the basic need is the same, namely to lay or build soundtracks (the sounds that go with the pictures) and then to mix them creatively to produce a cohesive soundtrack to counterpoint and expand on the pictures. In many discussions on sound reproduction the term 'quality' (that is, technical perfection) seems to be applied as the only criterion for judging

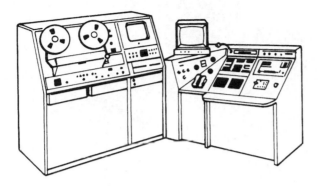

1950s
2″ videotape recorder,
black and white then
colour

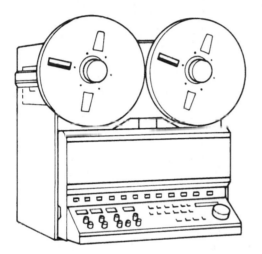

1970s
1″ videotape recorder

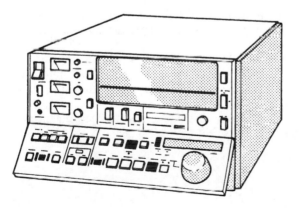

1980s
U-matic, the first
cassette format; now all
videotape recorders use
cassettes

Figure 1.3 The shrinking size of videotape equipment

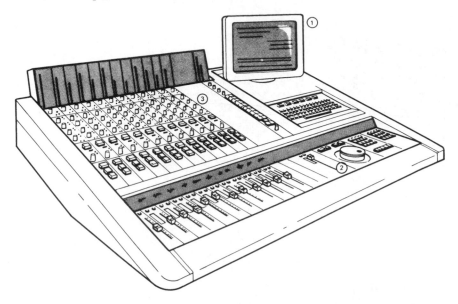

Figure 1.4 One of the first digital audio workstations (DAWS). 1, Workstation visual display screen; 2, workstation controller; 3, input channels

sound. However, discussions on visual reproduction use the term 'quality' as only a part of the overall judgement, for photography is often considered on artistic grounds as well as on technical quality. Likewise, we ought to consider the soundtrack in terms of artistic judgements, for sound is part of the whole creative work. In audio post-production it is important that the audio mixer has respect for both the pictures and the sound he or she controls, since there is an interaction between the two. The sound image and the vision must be cohesive; if they conflict the viewer will be lost.

The basic functions of the soundtrack are as follows:

- To enhance the pictures and to point up visual effects.
- To add three-dimensional perspective to the single-dimensional picture, even without surround sound.
- To enhance realism with sound effects and acoustics.
- To give a geographical location to pictures that may be visually unrelated.
- To add dramatic impact with music and effects, perhaps with counterpoint.
- To add contrast by changes in volume.
- To provide programme sound that is intelligible and easy to listen to, in every listening condition.

2

Video, film and pictures

The purpose of this book is to describe audio post-production techniques. However, to fully understand the post-production process a basic knowledge of picture recording and operations is necessary, and in this chapter we look at the basic technology behind video recording and photographic picture recording.

Moving images, whether they be stored on film or videotape, are nothing more than a series of stationary pictures, each slightly different from the previous one and each capturing a succession of movements in time. When these pictures are shown or projected at speed, the brain interprets them as continuous motion, a phenomenon known as persistence of vision.

It takes approximately a twentieth of a second or 50 ms (micro-seconds) to visually register a change in an image. When the images change at greater speeds than this, we see pictures that move.

Film

A film camera records pictures photographically. This is achieved by pulling down a ribbon of film using a claw which engages onto a sprocket and positions this emulsion stationary in front of the lens. A shutter operates for a fraction of a second to expose the image on the film and then, while the shutter is closed, the next frame of unexposed photographic film is positioned (pulled down) for the shutter to open again for exposure (Figure 2.1).

The image is reproduced in a projector and in order to reduce flicker each image is reproduced twice. Three international width formats are used, 16 mm and 35 mm for television use, and 35 mm and 70 mm for cinema projection; 70 mm prints are, in fact, shot using a 65 mm camera negative, and 16 mm is often exposed, particularly in Europe, using a larger than standard-sized aperture called Super 16 mm which is used for theatrical release and for 16 by 9 wide-screen television. Super 35 mm is a similar format. On both, the soundtrack area is 'incorporated' into the picture area offering an enlarged picture aperture.

The intermittent movement produced in a film camera at the picture aperture is unsuitable for sound recording, which requires continuous smooth

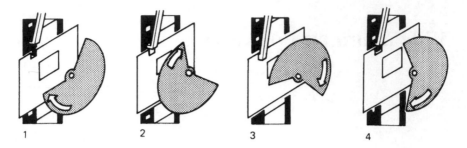

Figure 2.1 The claw and shutter mechanism of a film camera: 1, exposure of the film; 2, claw pull down; 3, film advanced; 4, shutter opens again *(From S. Bernstein, by permission)*

motion. Sound is, therefore, recorded separately on a synchronous audio recorder; this might be an analogue or more likely a digital tape format, a hard disk format or even a solid state recording device in synchronization with the picture (double system). At one time, for news operations, recordings were actually made on a magnetic track bonded to the side of the picture, at a point away from the picture where the intermittent motion of the claw could be smoothed out (above the picture-head in 35 mm and below it in 16 mm). This remains the standard for replaying release prints in a theatre, although the sound is recorded optically or digitally but not magnetically. Film normally runs at a speed of 24 frames per second in the USA and 24 or 25 frames per second in Europe.

Telecine

Today's budget requirements mean that film is often used only at the shooting stage; production work is then carried out in the more cost-effective video computer environment. This requires the film to be transferred to video. This is carried out using a telecine machine. Eventually it may be necessary to conform the computer video edit to the original camera negative so that a film print can be made from the final edit to be released in the cinema. Various methods are used to ensure that the original can be edited to match the computer edit.

Video

The television system works on a similar principle to that of film, using persistence of vision. However, the image is produced as an electrical waveform, so instead of focusing the image onto a photographic emulsion, the television camera focuses its image onto a light-sensitive charged coupled device (CCD) made up of many thousands of elements or pixels. Each of these picks up variations in light, and breaks the picture down into its separate elements. These are sent one after another, sequentially, to the transmitter or recorder.

To reproduce the picture (Figure 2.2), an electron beam which is 'controlled' by the camera is directed onto a phosphor screen that forms the front

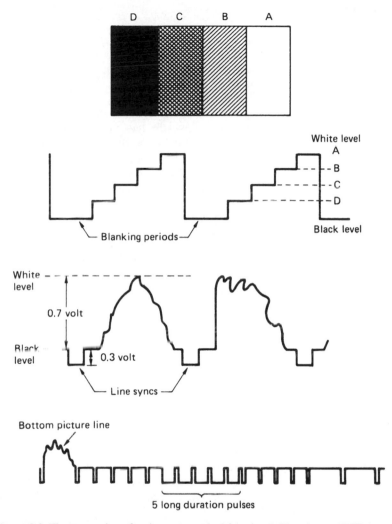

Figure 2.2 The conversion of a picture to a television signal. The picture ABCD is converted into various signal voltages. In the blanking period between the lines, the electron beam flies back; the sync pulses are added at the end of each field to keep the camera and receiver in step (*Courtesy of Robinson and Beard*)

of the television monitor, tracing a picture. The beam is muted during its 'fly back' journey from the top to the bottom of the screen in order to stop spurious lines being seen. The electron beam scans across the picture at a speed of 15 625 times per second. Each vertical scan is called a 'field'. On the European Pal 625-line system, 25 complete pictures are presented every second and the screen is scanned 50 times. In America, on the NTSC system, 525 lines are produced at close to 30 frames per second with scanning taking place 60 times. Since the picture has been split up by scanning, it is important that it is reassembled in step with the camera. Two pulses are sent to ensure this, one a line scan pulse and one a field scan pulse.

Video recording

A colour video signal has a band width of 5.5 MHz and specialist techniques are needed to record this signal. An ordinary analogue tape recorder such as a domestic cassette recorder can only reproduce a range of 10 octaves, whereas, to record video successfully, an 18-octave range is required. On our analogue tape machine this would require speeds of many miles per hour. The only practical way of achieving this necessary speed is by using a rotating head, which scans the tape at a high speed, while the tape itself runs at a slow speed to allow a reasonable amount of picture to be stored. A special control track is recorded to precisely define the position of the tracks written by the recording drum on the tape. This allows the head to be synchronized correctly to replay the recorded picture. Digital audio tape recorders (RDat) use this type of technology to allow their digital data streams to be successfully recorded. Digital video recorders use even more sophisticated technology, developing scanning systems further.

There are two forms of video signals, composite and component, used in video recording. Composite includes colour information along with the brightness or luminance information, and is found in the digital D2 format. In component video the colour information is separate from the luminance. Digital component is intended as a direct replacement for the older analogue formats. Analogue component systems include the traditional Beta SP format which is still widely used. Beta SX, D1 and DVCPro are examples of some of the other digital component formats.

The scanning head was first successfully used by Ampex in 1956, when it launched its two-inch quadruplex video recorder. In this format, four heads were equally spaced around the edge of the spinning head drum. This is mounted at right angles to the tape, one head taking over the recording when the previous head leaves the tape. Stationary images could not be viewed in this format. The linear speed is 15 inches per second but the effective recording speed is over 100 ft per second.

Video recording was developed further and the tape was scanned at a greater angle, allowing still images to be produced at standstill. Then to protect the tape it was placed in a cassette. These techniques form the basis of tape-driven recorders today.

In the 1970s videotape began to replace film in the broadcasting industry. A small format was developed that could match the quality and versatility of 16 mm film equipment in both size and quality. This was the helical scan U-matic three-quarter inch tape cassette system (Figure 2.3); but with a linear speed of only 3.75 inches/second the audio quality was poor, as was picture quality with a bandwidth of 2.2 Mhz compared with the 5.5 MHz required for a TV broadcast system. All this placed severe restrictions on the number of video generations that could be made in editing to produce the final product. The U-matic format was developed further but is now only a viewing format found in some audio post-production suites (Figure 2.4).

In the mid-1980s the Sony Betacam and the Matsushita/Panasonic MII camcorder video cassette systems were introduced, which revolutionized location shooting in Europe. They proved to be capable of matching and emulating high-quality 16 mm film production, even giving the elusive 'film look' so desired by film producers. The Beta format has been superseded by a

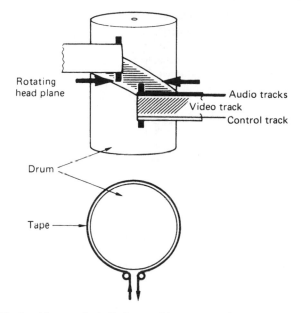

Figure 2.3 The head layout of a helical scan videotape recorder

variety of digital formats from various manufacturers, each of which aims to serve a particular part of the market at a particular cost.

Once there were a few standard video recording formats, now there are more than 10 altogether. Audio post-production suites have to be ready to accept any of the current formats used in the field. It is impossible for any one

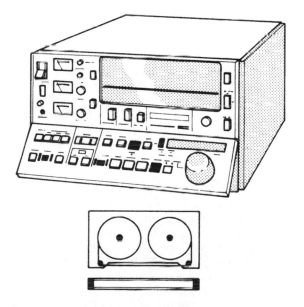

Figure 2.4 The U-matic format video cassette machine

Remote-1 in (9P) / Remote-1 out (9P)

Pin No.	Controlling Device	Controlled Device
1	Frame Ground	Frame Ground
2	Receive A	Transmit A
3	Transmit B	Receive B
4	Transmit Common	Receive Common
5	Spare	Spare
6	Receive Common	Transmit Common
7	Receive B	Transmit B
8	Transmit A	Receive A
9	Frame Ground	Frame Ground

External view

⑤ ④ ③ ② ①
⑨ ⑧ ⑦ ⑥

Remote in (9P)

⑤ ④ ③ ② ①
⑨ ⑧ ⑦ ⑥

Remote out (9P)

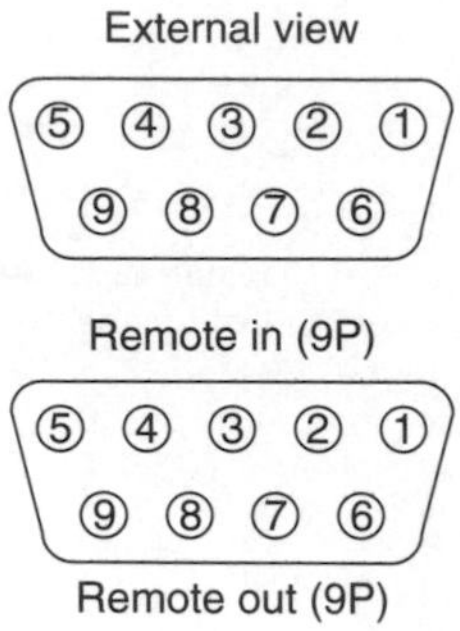

Figure 2.5 Sony nine-pin connection used in many editing interfaces, picture and sound

suite to provide machines in all the formats required (although some are compatible in one way or another), so these are often hired as needed. They must
be able to interface with equipment already available, from the audio
(digital/analogue), picture, synchronization and transport control point of
view, so it is important to check this. Most video recorders follow the Sony
P2 protocol for interfacing with editing suite controls using a nine-pin socket
(Figure 2.5).

Of the wide variety of formats now in use, most are digital and of those
the majority use compressed video. Much of the original thought on video
compression comes from MPEG, the Motion Picture Experts Group, whose
MPEG II format is widely used. Compression allows less tape to be
used in a recording, with resultant cost savings. Compression means throw-
ing away some of the information that is not really needed and reconstructing
it later with hopefully hardly any discernible picture or sound loss.
Compression can range from 10:1 for Sony SX to a mild 2:1 for Digital Beta
but the resultant picture quality after compression mainly depends on the
amount of processing power of the computers rather than the actual methods
used. The actual digitizing of a colour video signal requires sampling
the luminance (the brightness) and the chrominance (colour). In the D1,
D5, Digital Betacam, Betacam SX, Digital S and DVCPRO 5O systems, the
luminance is sampled four times to each chroma signal twice, hence the
term 4:2:2 when referring to the type of digitization. DVCPRO and
NTSC DV reduce the chroma resolution by sampling at 4:1:1. DVCAM in
PAL samples at 4:2:0. Digital video computer editing systems producing
broadcast quality pictures often record using a 2:1 compression and 4.2.2
processing.

Modern digital video recorders offer all the audio facilities of analogue
machines but with the advantage of four channels of high-quality sound. A
high-quality digital audio channel requires a data rate much lower than that
needed for video. Altogether, adding the data to record audio onto a video
track increases the data rate by only a few percent. The audio can then be
recorded by the video heads as small blocks of audio placed at specific intervals alongside the main tracks. At the same time, the audio sampling rate is
locked to the video rate. Since picture recording is intermittent, the recording
does not take place in real time but is compressed and expanded, so covering

the time gap. As information comes out of the 'audio buffer' it immediately releases more space.

The format chosen for a production will depend on the type of production; for drama and digital cinematography Digital Beta might be chosen. For news, DigiBeta is too sophisticated and Digital S or DVCPRO might be more appropriate.

All the digital formats have one thing in common: they provide digital audio that is of high quality and that can be copied many times without quality losses. Sixteen-bit quantization with 48 kHz sampling is the standard.

Some videotape formats are compatible with the compression systems on disk, allowing the fast transfer of data into a workstation from tape. Computer disk systems allow instant access to both sound and video material.

Viewing

In the audio post-production process, the soundtracks are mixed and edited against a picture format that is the most appropriate to the suite. The formats available are not too significant; all that is important is that the format can be copied to a format used by the suite, traditionally the U-matic video cassette recorder (VCR). This does have restricted picture quality, but at a low outlay, options are available for full synchronizing time code control and standard interfaces are available. More recently the Beta SP format has found acceptance with its good-quality pictures and sound and at moderate cost. This format is still used as an exchange format for video programmes and most broadcasters and production companies possess them. Edited masters are sometimes used in audio post-production systems, and are most likely to be in the broadcast cassette format.

Unfortunately, video cassette recorders differ considerably in their mechanics and suitability for post-production interfacing.

Video computer disk recorders of various makes are also used to replay pictures for mixing to; they must provide exactly the same facilities as their tape-driven counterparts. Their quality depends on their cost; they provide instant access to located points within the programme material. Some audio workstations are able to record pictures as well as sound. These systems offer fewer interface problems and no incompatibility. However, the image quality is often a compromise, usually using MPEG or JPEG (Joint Photographic Experts Group) compression systems.

Viewing images

Pictures viewed in the audio post-production suite can be shown on a monitor, or projected via a video or film projector. Ideally they should try to re-create the final viewing and listening environment. Images must be easy to see, with good definition, and, if projected, well lit. Video pictures can be successfully projected onto a screen up to a size of about 12 feet by 8 feet although quality is very dependent on the viewing distance. The picture definition and colour rendition are often inferior to those of monitors. Although impressive, projectors do not re-create, for example, the standard domestic

TV environment, and the poor definition and reduced environmental light levels necessary to provide a bright picture can lead to eye-strain.

Television monitors are graded according to quality, with grade 1 being used for checking broadcast transmission quality. Monitors (or projection televisions) should be capable of replaying both NTSC and PAL pictures.

In the large film re-recording theatre, or dubbing theatre, a film projector may be needed to produce sufficient light output for the screen. It will be specially adapted to run both backwards and forwards and possibly at fast rewind speeds. This is, however, a compromise in design. Picture quality may suffer since fast winding speeds and high picture definition cannot be achieved in one machine.

To assist the re-recording mixer to mix the various soundtracks of a production accurately, a timing counter is displayed, either under the screen or within it, or on the computer workstation screen. This must be clear and easy to read. For film operations the timing counter often displays 35 mm film feet, whereas for video operations a code is 'burnt into' the picture, displaying time (Figure 2.6).

Comparing film and video

Video and film are very different technologies. Film is mechanically based and comparatively simple to understand, whereas video is dependent on complex electronics, but both systems have their own attractions.

Film has been established for over 50 years and is made to international standards accepted throughout the world. Indeed, it was once the only

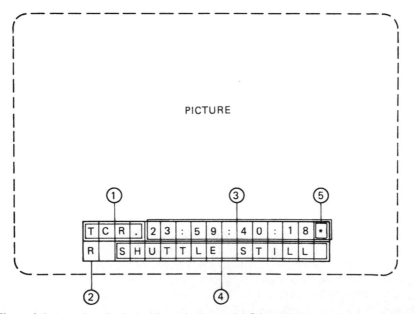

Figure 2.6 A monitor displaying 'burnt-in time code' ③ in addition to video data. ① Type of time data; ② videotape recorder assignment; ④ videotape recorder status; ⑤ indication of vertical interval time code

medium of programme exchange in television. This changed with the introduction of the videotape recorder, and in particular with the introduction of the broadcast video cassette format. Such machines are now being mass produced; more than 300 000 Sony Beta machines have been sold at prices lower than a film camera, and with minimal operating costs (film stock, processing and printing can be expensive). Film is now a specialist quality format, used in cinemas, and sometimes for television production. In 35 mm form on location its image is unsurpassed in contrast ratio and definition, and the film camera will be used in theatrical productions for some years to come. The projection of film images remains an excellent way of displaying images but high-definition video can produce comparable results which can be transmitted directly to a motion picture theatre.

3

Audio post-production equipment

The audio post-production process has radically changed since the 1980s. Computer technology has streamlined operations, reducing staffing levels and cutting equipment costs. Setting up a film mixing studio once cost perhaps five times more than a film camera; now the cost of one such camera will purchase the entire equivalent system in the form of a digital audio workstation. These workstations manipulate audio with exceptional ease in the digital domain, sound better and are as versatile as their analogue counterparts.

Because of the original high financial outlay, analogue equipment will stay in use for some time, particularly with institutions and broadcasters who are not affected by the need to follow fashion. Independent studios, however, must please their clients who often look for state of the art equipment. This can be hired out at a greater return. However, analogue equipment can be as effective as modern systems, particularly if run by competent and well-managed staff.

Digital workstations are at the heart of the majority of audio post-production systems, but there are other pieces of auxiliary equipment which are required as part of the system – video players and recorders, audio recorders, CD players and even sprocketed film recorders. These are often provided with specific options for audio post-production by the manufacturers. Common to all video audio post-production studios are video recorders.

Video recorders and sound

There have been over 30 different video formats since video recording began, all capable of recording sound in some manner. Ten or so of these may be met in productions today and require to be played off in the audio post-production suite. Some formats will need a 'hired in' video player, others will have to transferred at outside facilities houses onto a format readily available in the audio post-production suite. A list of formats is shown in Figure 3.1.

Videotape recorders are optimized for video recording, and in order to obtain a good video performance thin oxide coatings are used. This is quite acceptable for digital recording but unfortunately produces problems in analogue (longitudinal) audio recording formats with increased distortion at low

```
A selection of video formats

Format      Use                              Audio

Quad        Archive 2" tape                  1-2   tracks analogue

1" C        Archive 1" tape                  3-4 tracks analogue

U-matic     Industrial, various formats      2 tracks analogue

Betacam     Broadcast standard, various      2 analogue, 2 AFM

M11         Broadcast, various               2 analogue, 2 AFM

D1          Mastering studio (D)             4-track 20-bit PCM

D2          Mastering studio (D)             4-track 20-bit PCM

D3          Mastering studio (D)             4-track 20-bit PCM

D5          HDTV (D)                         10-track 20-bit PCM

DigiBeta    Origination drama etc. (D)       4-track 20-bit PCM

Digital S   News documentary (D)             4-track 16-bit PCM

DV          News (D)                         2-track 16-bit PCM

DVCPRO      News documentary (D)             2/4-track 16-bit PCM

DVCam       News (D)                         2/4-track 16-bit PCM

SVHS        Industrial                       2-track analogue

SX          General purpose (D) to           4-track 20-bit PCM
            replace standard broadcast
            analogue Betacam
```

Figure 3.1 List of formats. (D) indicates a digital video recording format with or without compression techniques

frequencies. Additionally, low writing speeds make recordings of high audio frequencies difficult too, and noise levels are only marginally acceptable! In some formats, audio noise reduction is applied to both tracks to overcome these problems.

In addition, audio FM tracks of excellent quality are available on the traditional analogue broadcast format, Beta SP.

Audio frequency modulation recording

Originally, frequency modulation recording was developed to offer a means of recording high-quality sound on domestic videotape recorders. This system is still current, and although not digital, it nevertheless produces a sound quality far better than its analogue counterparts. In AFM recording the analogue signal is modulated, at a set frequency, onto a carrier wave, deviating up and down from its basic standing frequency. The number of times it deviates per second is the modulation frequency, and the degree of deviation is the modulation amplitude. The recorded FM signal is not as subject to distortion or noise as a standard analogue recording. Nor is the frequency response dependent on the head or tape losses. The signal also tends to go into compression on overload, rather than violent clipping distortion – the problem with digital recording. To work well, systems have to be carefully set up, particularly if video interference is not to be apparent in playback. This is often a problem with domestic AFM recording.

AFM tracks tend to be used on original source recordings, since they have only limited use in audio post-production and editing. Since, in AFM recording, sound can only be recorded when the picture is being recorded, audio cannot be extended beyond the length of the recorded picture, so when the picture is cut to another shot the sound has to cut with it. Thus audio 'longer' than the picture shot (that is, the heads and tails or handles) cannot be laid down, but analogue longitudinal recording has no such problems since the sound system is entirely independent of the picture.

The introduction of digital recording dramatically increased sound quality on video recorders. Now it is no longer necessary to record audio using analogue techniques – although these techniques still find use in some formats for cue use. The problems of playing off digital sound at high speed in a recognizable form can be overcome by recording analogue sound on special cue channels.

Sprocketed magnetic film, 16 mm and 35 mm

Traditionally, in the film audio post-production suite many separate pieces of sprocketed magnetic sound film are synchronized together, each soundtrack being laced onto a separate sprocketed magnetic machine or dubber (Figure 3.2). Sprocketed magnetic film is oxide-coated film of the same dimensions as camera film. Sound is recorded in the analogue domain. Many sprocketed magnetic film machines are driven together in synchronization by a low voltage (bi-phase) generator. This is the double system method where there are separate soundtracks and a separate picture. Although this was once a standard method of producing soundtracks to pictures, computer technology has radically changed these techniques. However, much archive material is held in this form for television use and sprocketed machines can often be found in telecine suites.

Sprocket drive film machines or dubbers

Film is a thick and unsupple material, which is difficult to drive smoothly and accurately using the sprockets. Sprung mechanical compliance arms and dampers are used to maintain tension over the recording heads, while large flywheels reduce 'wow' and 'flutter'. Originally, all magnetic film machines used sprockets to physically move the stock across the heads in a 'closed loop' drive. The latest generation of film machines use standard servo control systems which sense speed from a sprocket driven by the machine which monitors speed and adjusts it against an incoming reference which is also fed to other machines in the system (Figure 3.3). The spooling motors continually adjust the film speed to synchronize with the incoming signals. The film is driven by the spooling motors from the control circuits and nothing else – this is called a zero loop drive.

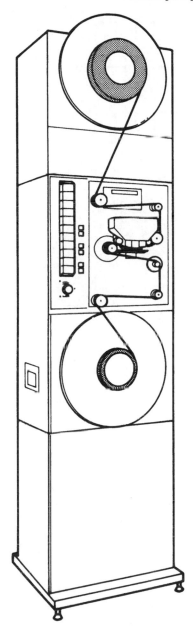

Figure 3.2 A sprocketed magnetic film machine using a capstan drive without a pinchwheel

Sixteen millimetre sprocketed magnetic film is now only found as an archive format. In 35 mm the soundtrack is usually positioned on the left-hand side of the film, in the same position as the soundtrack for a final theatrical release print. At 24 frames per second the film covers 18 inches in one second. Up to three tracks can be recorded across the film in normal usage,

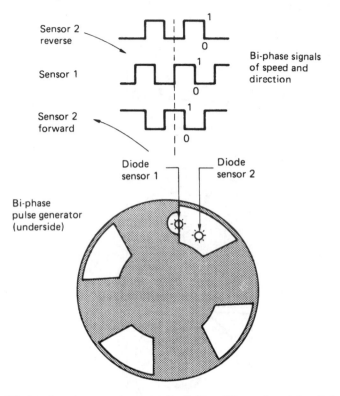

Figure 3.3 A tachometer generator containing light emitting and receiving diode sensors. This type of generator is used to provide bi-phase speed and direction pulses to control and synchronize sprocketed film transports and can also provide tachometer pulses if fitted to audio tape recorders for autolocating

although it is possible to record four or even six tracks (Figure 3.4). Modern film mixing facilities may require more tracks than these formats can handle, and small modular eight-track digital machines (dubbers) have been developed that can directly replace analogue film sprocketed machines at a fraction of their cost. The 70 mm motion picture film is rarely made, except for a very few high-cost productions, and only a handful of studios in the world are equipped to deal with it. Picture is handled in 65 mm widths before being printed up for release in 70 mm. Digital multitrack dubbers are sufficiently versatile to replace the four- and six-track 35 mm sprocketed machines in the re-recording theatre for multi-channel recording.

Requirements of sprocketed film machines and modular digital dubbers

- They must follow bi-phase signals, running up and down to speed smoothly, and stabilize without sound judder or loss of signal.
- There must be easy methods to check recording and replay synchronization.

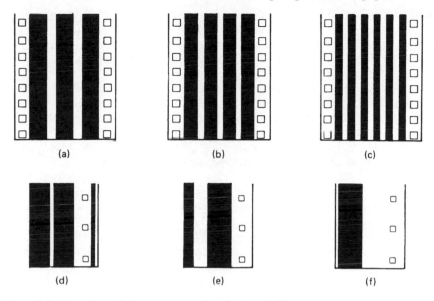

Figure 3.4 Magnetic track layouts on sprocketed magnetic film
35 mm   16 mm
a 3 track   d EBU 2 track stereo, time code
b 4 track   e 2 track edge and centre
c 6 track   f Edge track only

- They should be capable of being moved remotely by frames or feet (slip).
- They must be able to run in reverse with audio (more difficult in digital).
- They should be capable of producing or running loops for atmospheres, sound effects etc.
- They must be capable of running stereo soundtracks without phase problems (a particular problem with wide 35 mm formats).
- They should be able to accept a large variety of external synchronous signals.

Projectors

To view film in the audio post-production suite for theatrical presentations a film projector is often used. These projectors can produce very bright, high-definition, large images simulating a theatrical presentation in a large cinema. However, such machines are slow in operation compared to a video recorder and encourage a more leisurely style of operating. Significantly, equipment running in synchronization with the projector must also follow the speed of the projector as it is driven forwards and backwards up to play speed.

In many situations, modern video data projectors are capable of producing excellent pictures for large-scale production needs.

The projector must possess some additional requirements, beyond that of a standard cinema projector, in order to interface successfully with an audio post-production suite:

- It must be compatible with the sound system in terms of run up, interlock, maximum rewind speeds and the control system.
- It should not scratch or damage film, even at high speed. Although original camera film is never used, an expensive dubbing print or copy can easily be damaged in badly designed projectors.
- It should have a still-frame facility with a visible image, and in addition should be capable of showing an image at the highest speeds.
- It should produce the sharpest, brightest, flicker-free image possible.

It is impossible for one machine to meet all these criteria, and two types of projectors are available for film post-production operations:

1 Those which are adapted sound film transports with prismatic or mirror projection; these machines are capable of very high speeds at interlock but produce poorer quality pictures.
2 Those which use a cinema projection-type mechanism and are only capable of running a few times normal speed, but give excellent quality pictures.

Audio recorders and synchronization

At one time, audio recorders used in audio post-production, needed primarily to exhibit low noise characteristics to prevent the build-up of background noise from multiple copying, during soundtrack building and mixing. With the advent of digital recording these problems have been eliminated, although there are other problems, which can be overcome, such as the loss of sound at high and slow speeds and the complete loss of usable sound if overload takes place.

Invariably audio recorders are synchronized to their master video recorder by an electronic system called a time code (see Chapter 5). This code is recorded onto a dedicated track. Originally audio recorders needed to use one of their audio tracks as a time code track, reducing the number of soundtracks available on the machine. This was a particular problem with a stereo recorder which could not, of course, be used in stereo if one of the tracks was designated to record the time code (in fact, a spare track should be left between the time code track and the audio to prevent spillage of the time code noise into the audio). In the USA the half-inch four-track format provides an answer with audio on tracks 1 and 2 and synchronizing information on track 4.

This problem was resolved with the development of digital formats where the time code is integral within the system. Here the digital control data signal is used to help provide accurate time code synchronization. Manufacturers, aware of the needs of audio post-production, often provide options for their equipment to be interfaced with the time code.

Analogue recorders

Analogue multitrack, half-inch and quarter-inch analogue tape recorders are still found in audio post-production. Multitracks are used as part of a multiple track play-out and for music recording, and quarter-inch machines to

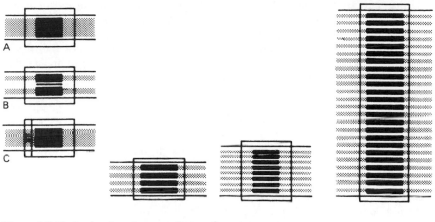

Figure 3.5 Professional analogue audio tape formats

Quarter-inch tape	Half-inch tape	One-inch tape	Two-inch tape
A Full track	4 track	8 track	24 track
B 2 track t/c	2 track signals	track 8 t/c	track 24 t/c
C neo pilot with separate push-pull	1 track t/c		
recording head (for film			
synchronization, see Figure 7.2)			

replay tapes recorded on location. Here they are found as a backup to digital, being a more robust format from both an audio recording and a physical point of view. A digital recorder is unlikely to work once it has been dropped, whereas an analogue machine will still often record something.

The various analogue professional tape audio formats are shown in Figure 3.5.

Noise reduction systems are often used in analogue recording to reduce residual background noise. Analogue noise reduction systems are unnecessary in digital recording and in fact can create problems with digital encoding if used. Noise reduction is used in some digital encoding systems such as Dolby's AC3 digital system.

Analogue noise reduction systems

Noise reduction systems work by compressing and expanding recorded sound. The overall dynamic range of the tape is restricted to a tightly controlled range during recording, so that low levels of sound are recorded at a higher level. On replay, the compressed signals are expanded up to their precise original dynamics, but now with improved background noise level. Noise reduction systems can suffer from disturbing problems of 'breathing' or 'pumping', caused when the replayed background noise from the tape is heard, varying in level, in quiet passages before and after the audio signal.

Generally speaking, most of the defects in 'compansion' systems depend on the degree of compression used. Dropouts can also become more obvious with compansion.

Noise reduction is regularly used in analogue recording for domestic audio cassettes, the analogue longitudinal tracks on Beta SP and in optical recording systems for the theatrical release of films. Although reducing noise, these systems do not reduce wow and flutter (i.e. speed variations) nor the distortion inherent in analogue recordings. Digital recording will eventually lead to the redundancy of noise reduction systems.

Digital audio recorders for audio post-production

Within audio post-production three forms of digital recording device are found, tape recorders, disc recorders and solid state recorders. These all exhibit the low inherent noise and excellent recording quality needed. Recording times vary considerably. They are all able to synchronize but with differing degrees of sophistication and at varying cost. Some are more robust than others. The basic digital recording system is shown in Figure 3.6.

All professional machines are able to run at a precise synchronous speed, allowing synchronization to other equipment. However, the accuracy of the synchronization over a period of time will depend on the sophistication of the crystal that controls time – just like a digital wristwatch. Accuracy may vary from a few minutes to many hours. Most of these machines can be locked to the master sync generator of a studio system, allowing precise lock to all other equipment if not to the outside world.

Most professional machines have available their own time code synchronizing boards, allowing them to follow other machines precisely with various possible interfaces ranging perhaps from bi-phase to the standard Sony editing interface P2 used to control many videotape recorders. Some will produce sound in all modes of operation from normal synchronization speed to standstill; others will not.

All digital machines need a buffer or random access memory (RAM) to allow them to deliver their recorded signals at the correct speed when the data stream is replayed from the recorded tape (Figure 3.7). The buffer acts in effect as a reservoir. The size of this RAM reservoir will affect the facilities offered and also the price. With lots of RAM a machine can, for example, scrub and slip sound, start instantly and vary speed. In addition, digital data errors can be dramatically corrected to reconstitute sound; in fact, sound unusable on one machine will play successfully on another. A machine with a large RAM is a useful asset to an audio post-production suite.

Digital equipment is continually being updated as research continues. Some equipment manufacturers offer updates to their equipment and software without cost; many offer a telephone help line and monitor customers' problems. Some publish target dates for providing software updates which can be downloaded from the manufacturer's Web site by those in the know and with the relevant know-how. This may well improve the performance of an early machine (designated version one – VO.1), but unfortunately an early purchaser may find that he or she is carrying out the research and development for the manufacturer. Such a purchaser may be left behind when a new updated big brother arrives, costing perhaps less than the cost of the original machine, and which now has far more facilities. An upgrade may then be available from the manufacturer at more cost, or it may not!

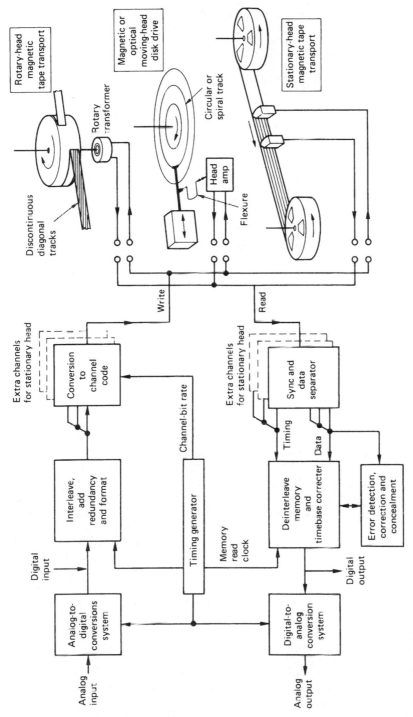

Figure 3.6 The basic digital recording system showing the rotary tape transport, the disc transport and the linear transport (*From J. Watkinson, The Art of Digital Audio, by permission*)

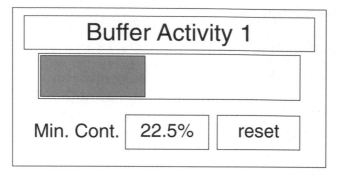

Figure 3.7 A workstation display showing buffer contents. Buffer memory must never fall below zero. The figure displays the lowest percentage of data in the buffer. The bar displays the current minimum level

Recording digital sound – audio levels

Digital audio is unlike analogue recording, for digital sound offers the ultimate in low noise and wide frequency response. Studios need to be quieter than ever and recording consoles set up properly to give the maximum dynamic range with minimum distortion on the gain settings. Ground loops need to be well sorted to reduce problems of low-level but audible hum between equipment. Most significant is the recording level.

The maximum recording level is at the top of the dynamic range; once over that point unacceptable distortion will result, so it is not uncommon for manufacturers to calibrate their recorders to record at –14 dB or even –20 dB below this level as an arbitrary '0 dB'. This does, however, leave some of the dynamic range of the system unexploited and there is a tendency for professionals to calibrate equipment with maximum headroom at peak level, leaving perhaps a dangerous figure of 0 dB headroom left! Like all systems, experience will tell all. Ideally there should be a standard recommended signal coding level so that recordings can be easily interchanged between studios, and there are recommendations, for example, from the EBU. Here it is recommended that the recording level is 18 dB below the maximum possible level. This applies in particular to R-Dat recordings.

Since there are no problems with head alignment on digital recordings, 10 kHz tones are unnecessary at the start of a tape, although a reference level is important together with a time reference indicating the loudest audio section, so that a check can be kept that there is no clipping introduced on copying.

A workstation graphic display for calibrating recording levels is shown in Figure 3.8.

Digital sound is:

- expensive in some professional formats;
- of almost perfect quality;
- available cheaply in domestic formats;
- excellent for access;
- subject to very unacceptable distortion if it reaches any overload point;

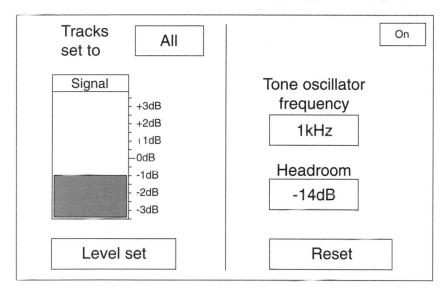

Figure 3.8 A workstation graphic display for calibrating recording levels

- subject to slow run-up times in decoding of sound if the system processors are fully loaded;
- without residual level after erasure;
- without a universal format;
- often unable to be checked in playback while recording, to ensure that the recording is present;
- liable to a complete loss of the recording, on disc sytems, if a fault develops.

4

Recording digital audio

The principle and practices of digital audio recording have little or nothing to do with past traditions of analogue audio. They are much more akin to today's computer technology and data storage.

Unlike analogue recording where the actual recording medium, the magnetic tape, limits sound quality, there are no limitations within digital audio. In digital audio the dynamic range is only controlled by the number of digital bits recorded. There is, theoretically, no degradation of the signal in copying, little modulation noise, and distortion is lower, no 'wow' and 'flutter', and no print-through. For the first time in the history of recording, almost perfect audio recordings can be made. The quality problems of digital recording are now, essentially, concerned with the conversion of the signal, from analogue to digital and from digital to analogue. However, once outside the studio, digital audio is likely to be compressed and encoded to allow use in transmission and consumer formats, and here there may be a quality problem.

Ideally, in audio post-production, recordings stay in the digital domain from R-Dat or solid state location recording through digital editing to digital transmission to the home or digital release in the cinema, finally emerging as analogue, so we can hear audio at the loudspeaker.

Digital signals require to be recorded at a much higher density than their analogue counterparts. No longer does the frequency response of the system merely need to reach 20 kHz where the signal is directly transferred as an audio wavelength to magnetic particles of 'miniature magnets'. Now the analogue signal is converted to ONs and OFFs but with a full audio bandwidth requiring to reach several megahertz, for the bandwidth for a digital recorder needs to be something in the region of 30 times greater than for an old-technology analogue recorder.

Digital signals

Digital signals take one of two possible values; in recorded form these will become equally saturated points of magnetization – one positive and one negative. Noise within the system can be completely ignored, providing that the two values have not been so degraded that they are lost in this background noise. In order to record sound digitally, three successive steps are needed to 'encode' the material (Figure 4.1). The incoming signal is first sampled at regular intervals, and the magnitude is determined. These samples are then quantified by giving each a quantity value, from a set of spaced values, so that the original signal can be represented by a string of numbers.

Sampling rate

The sampling rate of a digital sound recording system is normally greater than twice the highest frequency to be recorded. In audio recording the sampling rate must exceed 40 Hz, but the audio frequencies outside half this bandwidth must be eliminated before coding takes place (aliasing) otherwise spurious audible signals will be heard in the final decoded sound. Half the sampling signal is called a 'Nyquist frequency'. It is best to consider the end application when choosing a sampling rate. In music recording the standard 44.1 kHz may be most suitable since this is compatible with the CD standard. For broadcasting 48 kHz is standard. Higher rates mean higher quality but less recording time. Conversion to other rates is possible, but best avoided. When the original analogue sound is sampled, the sampling must take place at precisely regular intervals, as any errors will introduce errors into the recording or reproduction. Unfortunately, dropout (a momentary loss in signal) is possible in magnetic recording and since this could destroy part of the continuous signal of pulses, the system must be designed to check itself (rather as vertical interval time code restores and checks itself on a videotape recorder).

Quantizing level

It is the quantizing process that determines the signal to noise ratio of the system or the depth of the noise of the system (the quality of noise is very different from that of the analogue system). It is possible, in certain very critical conditions, to hear the background level 'falling away' when the base or 'floor' of the dynamic range is reached. The minimum quantizing rate for professional results is considered to be 16 bits, with 24 bits reaching maximum quality.

In the third and final step, the numbers are converted to form binary digits, like time code. By these three steps, the original signal can now be represented by a series of digits. Each sample is called a word, and in the example shown has three binary digital bits. The recording consists of a string of data bits.

PCM – pulse code modulation

Although digital data is only a series of ones and zeros, decoding it successfully can create problems – a series of 1s might be mistaken for just one 1! To

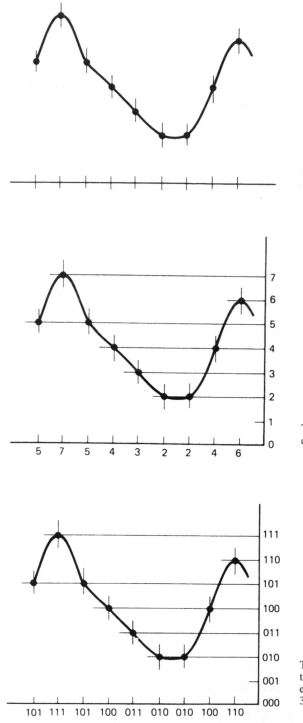

An analogue signal is
sampled at equal intervals

The sample signal is then
quantized

The quantized signal is
then digitized, in this
example using a simple
3-bit system

Figure 4.1 The principle of digital recording

make sure that this doesn't happen, the data needs to be referenced against a timing signal, so the bits are always correctly counted. This is accomplished by encoding the data onto another signal providing the timing information. PCM uses a steady pulse running at the sample frequency. The code is in fact pulsing, changing the starting time and width of the pulse according to the digital value of a word. Using PCM it is even possible to encode stereo into left and right channels, alternating the signals as packets of data; this is used in the AES/EBU encoding system.

There are certain techniques used to help record digital sound.

Spooling files

In audio post-production it is often necessary to hear audio at higher than normal speeds to find cues quickly. By simply increasing the digital system's clock frequency the drive will run faster. This will only work at a little above normal speed, and in order to go still faster, special files of spooling audio are memorized when the original recordings are made.

Dither

If analogue audio is sent to a digital recorder at a very low level, it is possible that it will not be processed properly – for it may fall below the quantization level, creating low-level distortion.

To reduce this effect, 'dither noise' is added as the signal enters the digital system. This constant noise means that the signal never reaches its 'floor' or base quantization point, helping to produce an acceptable sound but slightly reducing the dynamic range.

Oversampling

Oversampling reduces the amount that the filters operating in the analogue to digital conversion are expected to do. It works by adding extra samples to the bit stream and then raising the sampling rate to accommodate this. When the signal is reconstructed at its normal sampling rate, the extra samples are removed.

Error correction

In any system of recording, errors can be expected – a particle of dust on a tape can result in dropout of the signal. The high density of magnetic data recording on tape means that this will result in a loss of signal or a click. To eliminate this, an error correction system is used. Simply, this consists of recording the same digital data two or three times, but this takes up recording space and more often 'parity checks' are made. Here an additional bit is sent with each 16-bit word to 'monitor' and identify corrupt data. Interleaving offers another alternative – here the data is recorded out of order so that whole words are not obliterated, only parts. Recording with mechanical transports will always introduce errors, more with tape than discs; error corrections can eliminate these problems, allowing many generations to be made.

Memories

To record digital sound a recording system or a memory is required. This could be a spool of tape, a semiconductor integrated circuit – motionless recording – or a magnetic computer disc. The most common form of memory is the disc.

Digital recording on disc

Magnetic computer discs can record large quantities of data. They provide an answer to the needs of storing digital audio, offering good storage capacity and excellent access to the recorded information. Two forms of disc are available, floppy discs and hard discs. In floppy discs the magnetic recording medium is flexible, they are cheap to manufacture but have a very small capacity (100 kilobytes upwards to a few megabytes). Their use is limited to recording only small samples of audio, as well as filing information, such as edit lists and console set-up details. On the other hand, the hard computer disc has excellent storage capacity and systems are available that can record hours of magneti- cally recorded digital sound, with almost instant access. The circular magnetic tracks on the disc's surface repeatedly present data to the replay head. The rota- tional speed of the disc and the head seek time determine the access time. Audio is recorded in a discontinuous nature on disc and during replay it is stored and then released within the random access memory (or buffer) as an uninterrupted stream of digital data at the selected sample rate. A single disc drive can access several different tracks at once and can appear to have many tracks playing at the same time. Since there is no fixed physical relationship between the differ- ent tracks, it is easy to move the tracks relative to each other. The buffer also helps in synchronization to ensure the data stream locks with time code.

Buffer activity can be displayed in workstations, showing how much digital audio data is in the internal buffer memory at any instant.

The buffer is continuously refilled as data is read in bursts from the hard drive. The digital audio will then be played out directly from the buffer at the selected sample rate. To maintain the correct timing of the tracks the buffer, like a water reservoir, must never fall below zero. The digital signal proces- sors try to ensure this never happens. If systems become overstretched they may stop recording or replaying; this can be solved by reducing the amount of work that the disc head is asked to do by merging some of the operations, reducing the number of tracks or using a lower sampling rate. Replacing the hard drive with a faster one is the long-term solution.

Personal computer-based systems

Personal computers provide the basis or platform for many manufacturers' sound recording software, allowing an office or home computer to become an audio recorder. Information to control the recorder appears on the screen from the computer's software. Initially this seems a budget-conscious option. However, such systems in audio post-production not only need to record at sufficiently high quality, a minimum of 48 kHz sampling at 16-bit quantization, but must also synchronize successfully and be capable of interfacing for remote operation with other transports. Digital inputs and outputs may also be needed. This can make the software expensive for such a simple operation. Calculating the real cost of a system may be difficult as there are so many variables.

Packages relying solely on the host central processor unit will often require the latest model and perhaps even more power than the version available – a costly upgrade may be necessary. The number of channels the system supports will depend on the size and number of disc drives. Some packages don't use the central processor as a host but have their own hardware instead, which increases cost. Cards have to be physically plugged into the motherboard of the CPU which must have space available. If problems occur with such systems, manufacturers may well blame other parts of the system, not their own. Viruses can be found in an existing hard disc that can also create problems with a manufacturer's software.

To reduce problems, many manufacturers offer complete working systems, including a host computer. Should there be problems with such systems, it is up to the manufacturer to sort them out. The more sophisticated the system, the more viable it becomes to purchase a system complete with platform – which is becoming proportionally less of the cost. Generally, platforms are decreasing in price at a greater rate than software. A desktop computer-based workstation is shown in Figure 4.2.

The standard personal computer or PC originally developed by IBM and the Apple Macintosh computer can provide the basis of an audio recorder – usually supporting two or more tracks. Much of the systems' complexities relate to the control of the system, processing, synchronization and interfac-

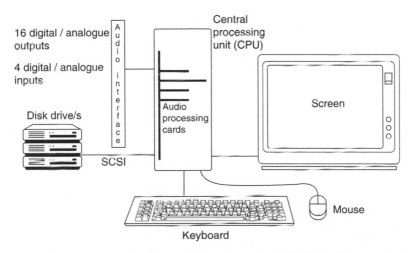

Figure 4.2 A desktop computer-based audio workstation. SCSI: Small Computer Systems Interface

ing generally rather than the number of the tracks offered; these systems are therefore most frequently offered as multitrack digital recorders or, in more sophisticated form, workstations manipulating sound for editing and mixing – the core of audio post-production systems.

Workstations

Workstations are complete audio post-production systems, often capable of controlling outboard equipment, synchronizing it and editing and mixing sound within the digital domain. A workstation audio post-production system must not only have sufficient capacity to be able to record all the necessary material needed for a programme, but must also have fast processing power to allow the material to be quickly retrieved for editing. Facilities are improving all the time, and manufacturers regularly update equipment with new software. The basic control of the system comes from the hardware, which can also be updated. The digital data is recorded on computer hard discs (Figure 4.3).

Sample rate	MB per track-minute
22.05 kHz	2.52 MB
32.00 kHz	3.66 MB
44.10 kHz	5.05 MB
48.00 kHz	5.49 MB

Figure 4.3 Disc space required per minute of track storage (mono) at different sampling rates

Workstations tend to use the Winchester hard disc system because it provides particularly good access; the height of the head above the disc is reduced to a minimum in order to increase the capacity. The head can move to a point on the disc within less than 6 milliseconds, and movement to adjacent tracks on the disc will take only a fraction of a millisecond – a negligible time for audio processing purposes.

As a guide, about 1 MB of disc space stores 10 or so seconds of audio and 1 GB stores about 3 hours. These figures relate to uncompressed 16-bit audio at a sampling rate of 48 kHz.

A typical low-cost workstation system supporting picture and audio consists of:

- a fast and expandable computer;
- sound and possibly vision digitizer card;
- 16-bit audio card;
- video compression card;
- large and fast hard drives;

- large and fast hard drive controller;
- high transmission rate between storage and display.

The crucial element is the speed with which the data can be processed to produce acceptable picture and sound. Such is the quality of digital disc recording that it should no longer be necessary to choose between recording systems on the basis of technical quality, except in specialist cases. Other factors now come into play, such as 'user friendliness', capacity, speed of access and the ease of transferring data from one system to another.

To transfer material from one workstation to another can prove a problem with hard discs. The disc, disc head and positioners are in a sealed unit away from contamination and the disc cannot be removed from the disc package. However, the whole system can be removed and in size is little bigger than a video cassette. One alternative to removing a hard drive is to use magneto-optical discs which record using a high-power laser, but with less capacity. It is also possible to record data onto tape; although this is more of an archiving format, it is extensively used to transfer data between video and audio workstations. Ideally, data should be capable of being transferred at a speed faster than real time.

Tape and disc share the same magnetic properties. Thus they both benefit from the developments in heads, oxide and coding technology. Tape is a recognized backup in information technology systems and it will continue to develop both for data and video recording, keeping in step with computer disc technology. Disc recording sacrifices density to gain fast access (for example, in workstations) whereas tape sacrifices access time to gain high density (as for use in videotape recorders). It will always cost more to store data onto disc but if a disc fails it generally takes all the data with it. If a tape fails, the tape can be removed and read on another machine.

Other disc systems – MiniDisc

The MiniDisc (Figure 4.4) provides two-track recording with standard discs and four-track recording with MD data discs, often in small, affordable, compact studio packages. Sound is compressed by 5:1 but maintains good quality using ATRAC (Adaptive Transform Acoustic Coding) technology at an economical price.

It records at 44.1 kHz sampling capable of delivering 16-bit quantization. The system is Midi compatible and MTC (Midi Time Code) synchronization and Midi machine control (MMC) is available. Synchronous by nature, long-term stability will depend on the accuracy of crystal control. These machines can form the heart of a small audio post-production system.

Within the broadcast and audio post-production field, MiniDiscs have found use as replacements for the traditional NAB analogue cartridge machines; they are ideal for holding sound effects for cueing in effects when dubbing on the fly. A whole generation of equipment has grown up which serves this small specialist market. It is claimed that one million rewrite cycles can be achieved without problems.

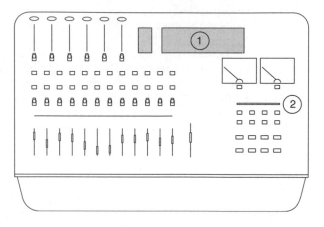

Figure 4.4 A MiniDisc system including recorder and console capable of Midi time code synchronization on a budget. 1, Graphics display screen; 2, MiniDisc slot

Compact discs

The digital compact disc found in the home was developed from video disc technology. Compact discs are marked with three letters to indicate the recording path (Figure 4.5).

Their instant access, immunity to handling problems and excellent quality make them ideal as a storage medium for sound effects and music libraries. They are also produced to an international standard, and can be recorded in a studio with specialist equipment at little cost.

The compact disc has a spiral track on its surface, starting at the inside of the disc and working outwards. The disc is 120 mm in diameter and has about 2000 information spirals in the programme area. The actual recording takes the form of a series of laser-cut indentations (pits). They are tied to a tracking speed of 1.2 metres per second, the rotational speed of the disc varying from 200 to 500 rpm to maintain a true linear speed. The digital data is recorded at a sampling rate of 44.1 kHz, with 16-bit resolution per channel.

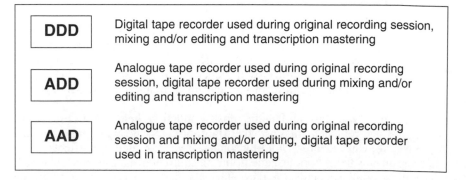

Figure 4.5 The markings used on compact discs detailing the recording path

The compact disc offers a high recording capacity and has the additional advantage of there being no physical contact, and therefore no wear, between the head and the medium.

Digital recording on tape

Audio magnetic tape has an excellent storage capacity, a large area of sound in terms of tracks, and can be rolled up into a small space. Access, however, is only possible by spooling through the tapes, which can be slow. It is impossible to gain access to the end of a tape without unrolling the material up to that point. Data discs, on the other hand, have their whole surface permanently accessible, but are not considered to be so robust and can fail with disastrous results.

The open-reel stationary-head digital audio recorder was introduced in the mid-1980s to replace the analogue recorder in the music recording studio. In the multitrack studio, the mixing console was far outperforming the tape recording system. In order to increase studio quality dramatically, it was merely necessary to replace the analogue tape recorder with a digital one. The Dash and Pro Digi format, reel-to-reel digital recorders were introduced, performing all the normal functions expected of an analogue multitrack recorder. They record sound by a system of multitrack heads, splitting data into various different channels. A robust, transportable machine with four recording tracks is available from the Nagra company, who are famous for their quarter-inch analogue portable machines.

Unfortunately, digital recording and open-reel transports are not entirely compatible. The short wavelengths required for digital recording rely on intimate head contact, and are very intolerant of tape contamination. To reduce these problems the rotating head digital audio tape format (R-Dat) was introduced using tape cassettes (Figure 4.6). These are small machines particularly suitable for location recording. However, even these can have recording problems, for their error correction systems can hide potential problems until it is too late, since less expensive machines may have limited data error warning systems. Ideally, error displays should indicate if error displays are recoverable, interpolatable or just non-concealable.

R-Dat

These transports are based on the highly reliable DAT data storage mechanics DDS. The portable machines are a standard for location recording. In the studio they perform well but operationally can be slow to synchronize to other pieces of equipment.

Originally offered for domestic use, the Rotary Digital Audio Tape format has been upgraded for professional use. Domestic machines may have problems in the audio post-production environment since they do not allow direct digital machine-to-machine recording at the 44.1 kHz sample rate. In addition, they are not locked to a precise speed. Professional machines offer sampling rates of 44.1 and 48 kHz with highest quality at 88.2 and 96 kHz. For twice the recording time, some machines offer a poorer quality audio at 32

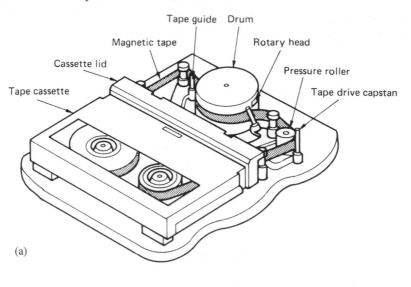

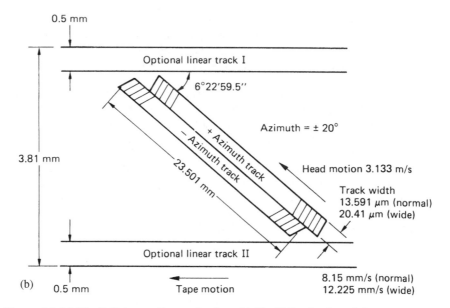

Figure 4.6 (a) The R-Dat recording mechanism. (b) The R-Dat head track layout *(From J. Watkinson, The Art of Digital Audio, by permission)*

kHz sampling and even four-channel recording. Up to two hours of recording can be made on one tape. Studio machines offer such facilities as scrub, varispeed and slip and seamless punching in and out of recordings. With tape recorders of the helical scan drum type, this is a complicated electronic process. The recording head has to follow a precise path with accurate timing as it enters record if no glitches or dropout is to be heard on replay. R-Dat records special automatic track-finding data within the audio track to help synchronize head to drum speed.

Professional time code synchronizing machines can cost as much as an industrial video recorder, but can offer a wide range of standard synchronizing facilities, including reading time code. The more sophisticated machines have four heads, allowing replaying while recording to confirm recordings have been made.

The track spacing on the drum scanning can be very close, allowing two hours of recording time at a sample rate of 44.1 kHz. Much of the technology for the recording system has been developed from the compact disc.

A tape width of 3.8 mm is used, matching that of an ordinary analogue compact cassette, although the cassette itself is slightly smaller. The tape is scanned by two (sometimes four) record/replay heads, giving a 90° wrap around the drum. The small degree of wrap reduces tape wear and the likelihood of tape breakage. The head drum is rotated at a speed of 2000 rpm. The normal tape speed is 8.150 mm per second, but one additional speed is available; there are four possible sampling rates. The speed itself is accurately controlled but, in domestic versions, insufficiently for synchronous recording. Domestic machines also include SCMS (Serial Copying Management System), a system that allows CD to DAT copies to be made in the digital domain but not then DAT to DAT; this is achieved by identifying a copying flag in the CD data stream (introduced because of the US Home Recording Rights Act). This severely restricts the usefulness of domestic R-Dat in the studio.

Access with R-Dat is good. The tape speed is slow, therefore rewinding to a given section is quick. The drum is small and lock-up immediate. There is a certain degree of user information that can be recorded on the machine, and cues can be inserted so the player will skip from one point to another for editing and quick retrieval. This facility is not designed, however, for great accuracy and it is not entirely suitable for time code information. The low linear speed of the cassette means that the longitudinal tracks of the R-Dat format are unusable for time code. However, certain manufacturers produce accurate time code interfaces and these machines provide an excellent time code tape recorder in the field. These machines have four heads, allowing replaying while recording, to confirm recordings have been made (confidence checking), a useful facility on location. Comprehensive error rate readouts should be included in the design to ensure that good recordings have been made.

Modular digital multitrack recorders (dubbers)

Standalone digital multitrack recorders were developed to replace their analogue counterparts in the music industry. They are small compared to their analogue partners, taking up only a few inches of rack space. Usually manufactured with eight simultaneous recording and replay channels, many machines can be interfaced together to record almost any number of tracks. Used in audio post-production they can be synchronized to systems to record digital sound for transferring material from one site to another. They are ideal for recording specialist multitrack formats such as surround sound or stems

for music and effects tracks (requiring six tracks). They can replace sprocketed magnetic film machines or dubbers interfacing with the standard film bi-
phrase synchronization system. Single-speed reverse and even looping of
sections in synchronization are offered. Displays can be in 35 mm film feet or
time; up to a few hours of recording time is available. Often higher quantizing rates above the standard 16 bits are available, but this reduces recording
time considerably. Some are able to copy machine to machine at high speed
with no loss of quality. On location as a compact recorder they offer the
opportunity to record many tracks for multiple music recording or multiple
recording of actors; they are usually mains powered. A modular digital multitrack recorder is shown in Figure 4.7.

Multitrack digital recorders are available in various formats. The first
modular recorders used the SVHS video cassette in the ADAT format,
running the tape on the VHS helical scan format at three times the normal
speed required since the format uses a standard analogue control track. A
standard 120-minute tape will only record 40 minutes of audio. These
machines are immensely popular for home music studio use and can be linked
up with others; they have full synchronizing facilities, often including Midi
interfaces. Unfortunately tape wear can be a problem and only a few hundred
passes are possible on one tape.

The Hi 8 format was the first modular multitrack format to find wide
acceptance in the audio post-production field for the interchange of programme material. This is based around the precision HI 8 video transport.
Machines offer full options for chasing time code and have their own integral
time code track. They offer 110 minutes of eight-track recording, more than
enough for specialized film use. Despite recording on tape with physically
fixed tracks, with their own RAM buffer they can offer varispeed and individual track delays as well as the usual gap-less insert recording. Sixteen separate machines can be locked up with the Sony P2 nine-pin interface. The
manufacturer uses a 25-pin single socket in and out for digital sound, offering eight tracks: the Tascam Digital InterFace (TDIF). The tape is enclosed in
the cassette even when recording and the control track, like DAT, is
Automatically Track Found (ATF). Here 108 minutes of audio are recorded
on the standard 120-minute video tape.

Hard disc modular multitrack recorders are able to offer far more facilities
than their tape format counterparts. Technically they follow the design of
audio workstations, but without the extensive editing facilities. There are

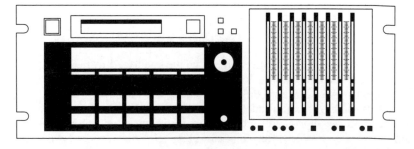

Figure 4.7 A modular digital multitrack recorder with eight bar graph meters, capable of
recording and replaying eight tracks at once

many competing systems on the market and they are able to offer all the facilities expected for audio post-production work. It is possible to provide immediate access, slipping of the individual tracks, reverse running at single speed and looping of sections in synchronization. Flags showing special start marks, first frame action and current video position are possible, with on-the-fly edits and displays showing film feet and time. Over three hours' recording time is possible on some machines and so are higher quantization and sampling rates. Some use removable Winchester discs and others removable computer data discs such as magneto-optical discs. All provide synchronization facilities and all are modular to increase the number of tracks available.

Motionless recording

It is possible to record and reproduce audio in a solid state device, with no moving parts. These devices find particular use in the field, where they are lightweight and shock resistant. The digital audio is recorded onto PCNIA flash cards similar to the ones found in laptop computers. They can record over four hours of mono sound using MPEG in audio compression mode, which provides quality equal to 16-bit 48 kHz sampling. All the usual non-linear facilities of editing and random access are available. Flash cards are in essence a vast number of capacitors interconnected by a framework of transistors addressed by a clock. Unlike normal RAM, they do not lose their capacitance when switched off.

Interfacing digital equipment

At the heart of an audio post-production suite is the workstation; here sound is recorded, manipulated and played off within the digital domain. Ideally, to maintain digital quality on a project, all audio connected with it needs to be recorded and processed digitally. The digital devices need to be of a quality better than or equal to that of the workstation. This usually means working with at least 16-bit quantization and 44.1 or 48 Hz sampling. Digital audio offers many different professional standards for interconnecting equipment, some of which carry the digital synchronizing information. Digital signals are sampled to a particular time structure; if two or more are combined it is essential that the time structures match. Synchronization needs to be more than just running at the same speed of, say, 48 kHz, because even infinitesimal speed variations may cause a low-level glitch or at worst no signal at all. It is usual to use a central high-quality synchronous generator to feed all digital equipment to maintain perfect synchronization. Many manufacturers provide separate optional boards to suit their customers' uses. Most common is the international AES/EBU interface which carries up to 20-bit audio data with synchronizing clock information. Two channels can be carried on the one cable. The standard type of XLR connector is often used with twin screen microphone cable.

This interface was designed to be used with balanced audio lines traditionally found in analogue studios. Other manufacturers' interfaces include, for example, MADI (AES 10) – Multi Machine Audio Interface (Sony, AMS

Neve, Solid State Logic, Mitsubishi) carrying 56 channels of audio, and TDIF (Tascam Digital InterFace) carrying eight in/out channels for Tascam's own make of small modular eight-track digital recorder. Some do require a separate cable carrying synchronization information. A unit for interfacing TDIF and MADI formats is shown in Figure 4.8.

To interface machines correctly requires some thought – the machines need the same interfaces, they must be set to digital as opposed to A/D conversion, they must be set to the same sampling rate and capable of the same quantization resolution (or set to conversion), and a check must be made see if a synchronizing (gen lock) signal is necessary; ideally a common synchronizing generator should be connected to the machines. Connections for audio will be XLR, phono or multiple D-type connectors. A video BNC socket may be used for synchronizing information. Unfortunately, not all equipment offers compatibility. If there is a connection problem between digital equipment then the opportunity will exist to use the analogue inputs rather than the digital ones. Professional equipment offers very high quality digital to analogue converters as an option to direct digital inputs. Using these will lose the very advantages of 'translucent' digital audio, since the converters, by their very nature, will degrade the sound quality; however, one transfer will produce hardly any perceptible change in quality. But the more the signal is removed in and out of the digital domain, the more problems there will be. Remember, however, that recording onto an analogue format will bring an appreciable loss in quality. These interfaces carry data in real time and are unlike computer data networks which carry data in both directions and in a packet format.

Special requirements for audio recorders

In addition to being able to record sound to the highest quality with exceptional reliability and in synchronization, an audio post-production audio recorder needs to perform certain other basic functions. It should, for example, ideally be possible to check that a successful recording is being made during recording. Tape-driven analogue or digital recorders and film recorders employ separate heads in order to replay sound during the recording to confirm that it has been recorded; this is called confidence checking. This is not possible with computer disc recording devices; however, these are exceptionally reliable, although when they fail a complete loss of information is likely with little chance of retrieval.

Confidence checking is essential on location shoots where the expense of having to reshoot may be prohibitive or even impossible. In studio operations it is less necessary to have confidence checking if there is continuous monitoring of the audio with spot checks at the end of the recording. This is a standard video recording practice where confidence checking is not possible.

In traditional audio post-production, the final soundtrack is produced by continually updating the audio mix until a satisfactory recording is made. The new mix may be created by re-recording over the previous one, in sections or scenes; the new section is inserted by punching into record mode at an appropriate point and re-recording the mix correctly. To do this successfully, recorders used in audio post-production must be able to switch into record silently when the 'punch in' to record is performed. In addition the 'punch in'

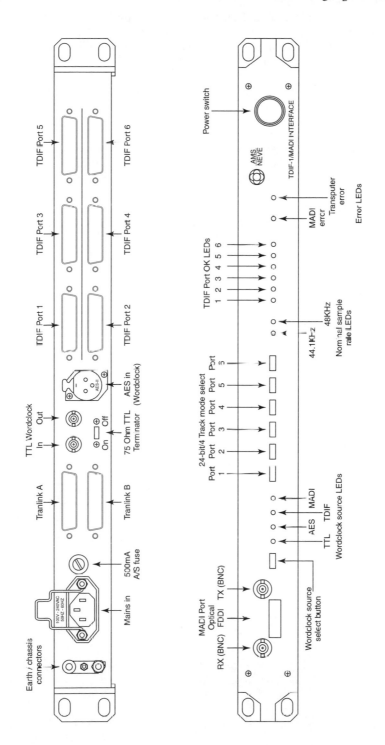

Figure 4.8 A unit for interfacing TDIF and MADI formats. Synchronization for the data can be selected from Wordclock, AES/EBU, MADI or TDIF (*Courtesy of AMS Neve Ltd*)

must be seamless and without glitches or it will be obvious that there has been an update. The same must apply to the 'punch out' of record; however, modern recording desks and workstations, with their ability to memorize mixes, can make this option less of a necessity.

Digital formats for audio post production need to be:

- of excellent quality;
- easy to interface with other equipment;
- easy to synchronize, ideally with their own dedicated time code track;
- capable of being checked for quality while recording, if tape driven;
- available with the correct quantization and sampling rate, e.g. 16 bit, 48 kHz;
- available with removable storage in a compatible form;
- offered with the correct standard digital audio interface, AES/EBU, TDIF, etc.

5

Synchronizing audio post-production equipment

An audio post-production suite is designed to bring together the various sounds that will make up a polished final production, allowing them to be accurately mixed together into the final audio track.

The sounds may be music, effects or dialogue but all must run in synchronization with the picture, and do so repeatedly, if the sound mixer is to be able to rehearse and refine his or her soundtrack mix to perfection. To achieve this degree of accurate synchronization, the conventional method was to use physically sprocketed film, or now to use the electronic synchronizing system called time code.

It is easy to grasp the simple mechanical principles behind film sprocket synchronization; sprocketed picture film and sprocketed magnetic film sound-tracks are locked together. A picture synchronizer demonstrates the system (Figure 5.1).

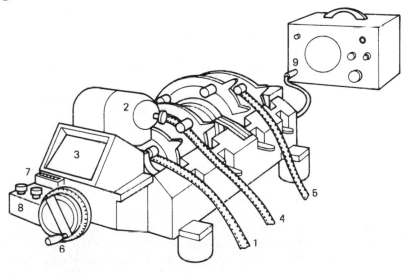

Figure 5.1 A synchronizer used for laying tracks. 1, Picture track; 2, light for picture; 3, screen; 4 & 5, magnetic sound tracks; 6, crankhandle; 7, footage counter; 8, volume controls; 9, loudspeaker unit *(Courtesy of A. Nesbitt)*

Various sprocket wheels, all of the same size, are locked onto a common shaft and driven at speed; each of the films engaging the sprocket wheel runs synchronously with the next. All run at the same speed and each is in synchronization with the next. It would be possible to adapt the synchronizer so that 35 mm film and 16 mm film were both running together on the same shaft. Both would be in synchronization – in 'lock' – but they would be running at different speeds. Now two different formats are running in synchronization together through a common synchronizing system. By connecting a counter to the sprocket wheel driving off the moving film, it is possible to time the passage of the film. If the film is run at 24 frames per second, the passing of 24 frames over the counter causes the timer to move one second. In this way frames, seconds, hours and minutes can be counted as the film passes – a form of time coding. But what happens if the recorded picture and soundtracks are not physically sprocketed? In this case an electronic sprocket is used, one originally developed for videotape editing.

Initially, videotape was edited by physically cutting the magnetic tape itself. However, this proved to be rather cumbersome since it was difficult to find a particular picture frame accurately on the electronic sprocket – the control track – which was used to synchronize the recorded picture with the revolving video drum. A poor splice resulted in the picture jumping – rather as though a bad film join had jumped in a projector gate.

Videotape editing techniques slowly improved and in the 1960s new electronic splicing systems were introduced; these systems did not require the tape to be physically cut but worked by electronically copying the picture from one machine to another, building up the edited material by re-recording the originals into a new, edited sequence. In order to achieve this accurately, a system was needed which enabled each individual videotape picture to be identified or labelled. This identification was in the form of time, on a clock, and had been developed by NASA to track missiles accurately. Today, it is possible to identify the frame to be edited, enter its time or time code into the edit controller, and sit back and watch the system automatically perform a precise edit, exactly as commanded. If the edit was previewed and thought unsatisfactory, it could be trimmed, at will, altering the time code in or out points.

Time code in video editing gives precise time references, allows compatibility for interchange of material, and allows video machines to be synchronized together precisely. It is this facility that is of particular importance to audio post-production, since video and audio recorders of any format need to be synchronized together by the code.

To achieve accurate synchronization it is necessary to have special comparator circuits in each piece of audio equipment. These circuits compare all the time code readings of the various machines and adjust their speeds to maintain synchronization. Unfortunately, until 1969 there was no standard code that could be used for synchronization, and each manufacturer had their own system. However, in 1969 it was decided to develop an international time code allowing interchangeability between systems. This code is called the Society of Motion Picture and Television Engineers/European Broadcasting Union code – abbreviated to the SMPTE/EBU code.

The SMPTE/EBU time code

Time code is recorded using a digital code. Originally used to identify frames on a videotape recorder, it can be equally useful in identifying an audio point on a tape or a hard disc. The code is accurate to at least one video frame in identifying a cue point for synchronization. The time span of a video frame depends on the speed of the system – not in inches per second but in frames per second. There can be from 24 to 30 video frames in one second. Once the code has counted frames it then counts seconds, and so on, identifying up to 24 hours of frames. The code then starts again. Each frame, therefore, has its own individual second, minute and hour identification. The basic time code structure is shown in Figure 5.2.

Each time code signal is divided into 80 pulses or bits, and each of these bits has only one of two states, on or off, referred to respectively as 'one' or 'zero'. The basic signal is a continuous stream of noughts – the signal that occurs as time code crosses midnight. However, if the signal changes its polarity halfway through a 'clock interval', the message represents a 'one'. This is a method of modulation known as bi-phase mark encoding. If listened to, it sounds like a machine-gun varying in speed as the time progresses. Distinct changes can be heard as minutes and hours change.

Each frame, therefore, consists of 80 parts. The speed of recording these bits will depend on the number of frames per second multiplied by 80. Thus, in the easy to understand European system there are 25 frames each second multiplied by 80, equalling 2000 clock rates per second. In the American system there are up to 30 (among others) frames per second multiplied by 80, giving 2400 clock periods per second. This rate of bit recording is known as the clock rate or bit rate.

These coded signals are used to represent the numbers recording clock time – the time code (Figure 5.3).

To view a time code display it is necessary to have eight separate digits. The largest number recorded is 23 hours, 59 minutes, 59 seconds and 29 frames (USA). To reach this specific time, all the numbers from 0 to 9 appear at some point, and have to be recorded. However, the code to record them is only in the form of zeros and ones.

To record numbers the code is divided into groups of four, each of the bits having only an 'on' or an 'off' state. Bit 1 of the group of four is designated numerical value 1 when switched on; Bit 2 is given the value 2; Bit 3 is given the value 4; Bit 4 is given the value 8 when switched on.

Table 5.1 shows the way the number values 1–15 are produced from a 4-bit code. For example, to represent value 6, Bit 1 is off, Bit 2 is on, Bit 4 is on, and Bit 8 is off. Bit 2 and Bit 4, which are switched on, add up to the numerical value 6. If the number 8 is required, only Bit 4 representing 8 is, of course, switched on. Obviously, if only the numbers 1 and 2 are required – such as units of hours on the clock – a group of only two bits is needed; not four bits which extends to 8 and beyond.

The table shows that using a group of four bits, any decimal number from 0 to 9 can be coded (in fact, the maximum number is 15).

The time information is spread through the 80 bit code – 0–3 is assigned to unit frames; 8–9 to tens of frames; 16–19 to units of seconds; 24–26 to tens of seconds; etc. Interspaced between the time information are eight groups of

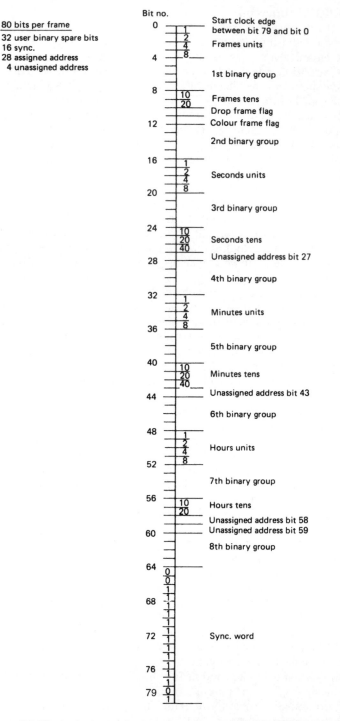

Bit no.

80 bits per frame
32 user binary spare bits
16 sync.
28 assigned address
  4 unassigned address

0 — Start clock edge
between bit 79 and bit 0
Frames units

1st binary group

Frames tens
Drop frame flag
Colour frame flag

2nd binary group

Seconds units

3rd binary group

Seconds tens
Unassigned address bit 27

4th binary group

Minutes units

5th binary group

Minutes tens
Unassigned address bit 43

6th binary group

Hours units

7th binary group

Hours tens
Unassigned address bit 58
Unassigned address bit 59
8th binary group

Sync. word

Figure 5.2 The basic time code structure as recommended in EBU Technical Standard N12 1986, Time and control codes for television tape recording *(Courtesy EBU)*

| Hours | Minutes | Seconds | Frames |

Figure 5.3 Time code read in the form of an HH:MM:SS.FF display

four bits each. These 26 bits are called user bits and to access their code special equipment is necessary. This information cannot be added once the code is recorded.

User bits record additional information such as video cassette number, the scene and take number, the identity of a camera in a multi-camera set-up, and so on. The user bit information is changed by re-programming it.

Other user bits are available to give standard information. Bit 11 gives colour recording information to prevent flashes on edits, Bit 27 is a field marker or phase correction bit, Bit 43 is a binary group flag, and Bit 10 gives specific time code type information. Bits 64–79 are always the same in every frame, since they tell the time code encoder when the end of frame is reached and also if the code is running forwards or backwards.

Time code and speed

Six different frame speeds are used in time code. It is possible to synchronize various different pieces of equipment together with different time code speeds, just as 16 mm and 35 mm film can be run in lock. However, this is complicated and requires specialist pieces of equipment. It is not to be recommended! Simplistically, time code requires continuity throughout a production. In Europe this is simple; in the US there are more options but the same principles, and here there needs to be from the start a clear understand-

Table 5.1 Chart of a four-bit code

Decimal number	Group of four consecutive bits, value when switched 'on'			
	1	2	4	8
0 =	0+	0+	0+	0
1 =	1+	0+	0+	0
2 =	0+	1+	0+	0
3 =	1+	1+	0+	0
4 =	0+	0+	1+	0
5 =	1+	0+	1+	0
6 =	0+	1+	1+	0
7 =	1+	1+	1+	0
8 =	0+	0+	0+	1
9 =	1+	0+	0+	1
(15 =	1+	1+	1+	1)

ing of how time code relates to each stage of the production as well as the destination of the final product.

24-frame time code (worldwide)

This code exists for film for theatrical release.

25-frame time code (PAL)

In Europe the television system operates at exactly 25 frames per second, and this speed is used in the production, distribution and editing of video material. This conforms to the European Broadcasting Union's PAL and SECAM systems. The speed is historically based on the use of a 50 cycle per second alternating current mains generating system.

Films for theatrical use produced in Europe are shot at 24 fps. In practice, the two are never mixed or only by mistake – problems do occur occasionally when film cameras running at 24 fps are used in television productions with audio recorders locked to 25 fps time code. Theatrical cinema films shot at 24 fps are usually shown on television at 25 fps which is the standard telecine speed – since each frame of film transfers exactly to each frame of video. At 24 fps this reduces the actual running time by 4 per cent and raises pitch, which can be returned to the original through a pitch shifter – a facility offered by some digital workstations.

30-frame time code (NTSC)

Originally this was used with NTSC black and white television. In America there is a 60-cycle generating system and this has been tied to a nominal frame rate of film at 24 frames per second. This speed was standardized when the sound film was introduced in 1928 (the slowest speed possible for apparent continuous motion). At its inception there were problems with transferring film to video. Film was shot at 24 fps and the system ran at 30 fps. The two are incompatible. To overcome this problem, additional frames are added as the film is transferred, periodically holding the film in the projector gate for three instead of two television fields (3:2 pulldown). This tends to give film a slightly jerky look. To produce better motion rendition, film can be shot at 30 fps which results in a one-to-one film-to-video frame ratio. On transfer to video, the film and separate sound have to be slowed down slightly (0.1 per cent) to accommodate the NTSC video system. This system is used for high-definition television, television commercials and music videos, but adds up to 25 per cent to costs.

The system is not relevant outside NTSC countries. It is used in high-definition television productions.

29.97 Non-drop code(NTSC)

In most non-video environments this SMPTE time code is related to the 60-cycle line frequency combined with a 30 fps code rate, as with domestic televisions.

With the development of NTSC colour systems, it became necessary to find ways to make the new colour broadcasts compatible with the existing black and white sets. The monochrome signal had to remain the same so an additional colour signal was added after the frame had been sent, simply by slowing down the video frame rate to 29.97. Thus clock time code and the TV signal run at slightly different times. This means that using a 30 fps time code system will result in a drift against an NTSC video recording which is 0.1 per cent slower than the nominal 30 fps speed. In fact the code gradually falls behind actual real time. The code is therefore modified to maintain speed – numbers have to be dropped out of the code. Over a period of 3600 seconds, i.e. one hour, there is a loss of 3.6 seconds (0.1 per cent). This is an error of 108 frames, and so 108 frames have to be dropped, through the hour, if real time is to be retained. In fact, real frames of video or film are not lost, only frame numbers are skipped to keep clock time.

To identify drop-frame time code, user Bit 10 is activated. On identifying this, the time code discards two frames every minute – except for every tenth minute; $60 \times 2 = 120$ frames per hour are admitted, except 0 10 20 30 40 50; $120 - 12 = 108$ frames per hour dropped. Code-reading devices will usually flag up which type of code is being read. SMPTE time code, in the drop-frame mode, therefore matches real time (within two frames a day). The lack of certain frames in the code can create further problems in operations. A synchronizer may be asked to park at a time that does not exist! Most synchronizers will deal automatically with this, however.

Drop frame is used in original video programming for broadcast use.

30-frame drop code(NTSC)

This finds limited use in film origination in the USA to improve motion rendition when the requirements are for television use.

Identification

Time code systems can seem even more tortuous when the time code frame rate and the picture frame rate do not match. Film can be shot at 24 fps and may have a 30 fps time code on the audio. Whatever the code chosen, continuity throughout the project is essential. For Europe there are few problems but if material is shipped overseas to the USA it must be well labelled (and vice versa) if there are not to be problems. The identification of the type of time code on material is most important, if systems are to synchronize in post-production.

Gearboxes

To interface various different time codes, gearbox devices allow changes of code. One code can be generated while another different code is generated in synchronization. This facility is standard in many digital audio workstations which use gearboxes internally; relating time code frame rates to the audio sampling rate is essential to problem-free digital audio.

Digital signal synchronization

In digital audio equipment there is a further time synchronization consideration since digital signals need to be very accurately timed if data is to be correctly analysed. If digital audio is to be combined successfully in any way, the signals must be synchronized to a common source. Even if the signals are out by 10 parts per million, audio problems can result. Time code creates a further burden on processors trying to synchronize audio and video devices.

Longitudinal time code (LTC)

In order to record time code, only a limited audio bandwidth is needed – something between 100 Hz and 10 kHz. This compares with a frequency response of 40 Hz to 18 kHz for high-quality audio recording. Although no special arrangements are usually made to record time code, there are problems. Often a specific track is dedicated to its use, particularly on video recorders and digital audio recorders. Recorded in the normal manner (like ordinary audio), this digital code is called longitudinal time code or LTC.

Recording time code

Analogue recorders find it difficult to record square waves, and strong, interfering, side band frequencies can be generated. This means that time code recording levels need to be chosen carefully, to minimize possible crosstalk or unwanted pickup into adjacent audio tracks. (Unfortunately, the time code waveform is at a frequency which is particularly audible.) Crosstalk can be reduced by recording the code at a lower level. However, if the level is decreased too much, the signal will itself be susceptible to crosstalk from the adjacent audio tracks, and from the background noise of the tape itself. Typical recording levels for time code are between –6 and –15 with respect to the zero or line-up level. A typical time code waveform is shown in Figure 5.4.

LTC can be recovered using standard techniques at speeds down to one-tenth of normal playing speed, or 2.5 frames per second. Unfortunately time code readings become unreliable as tape speed approaches zero; audio output

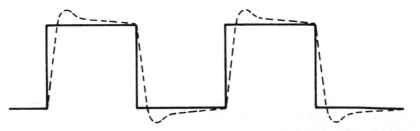

Figure 5.4 A typical time code waveform superimposed over the ideal theoretical waveform. The overshoot at the edges and the small amount of tilt is caused by phase response errors

is proportional to speed, so as speed reduces to zero, output also reduces to zero. These problems are overcome in digital workstations where time code is recorded as part of the overall timing of the system and time code can be accurate down to a part of a frame.

Replaying time code at high speed from a longitudinal track has problems too at high spooling speeds: with the tape in contact with the replay head, special amplifiers with an extended frequency response are needed. To recover time code at up to 50 times normal speed, the replay amplifier must have a frequency response extending to 100 kHz. Often, to eliminate these problems a recorder may generate pseudo time code from its time counter to simulate code in high speed – this only works with continuous code. It is unusual nowadays to encounter an analogue audio recorder requiring these techniques. Analogue tape machines are rarely used in post-production, and then usually only for loading a workstation.

Luckily, formats today have their own dedicated time code channels with set line inputs which do not need adjusting, and problems are rare if the manufacturer's recommendations are followed. They can also provide time code at varying speeds from 0 to 10 times.

Video recorders provide continuous time code from the LTC head and offer the facility to change to special vertical interval time code (VITC) (recorded within the picture lines) at low speed and tachometer roller when at high-speed running. Operating in auto they swap between these sources as required, using longitudinal time code as the correct code and VITC for timing. The correct switchings should be checked on a machine before use.

In digital equipment, time code will be recorded within the device on its own dedicated tracks. Time code is in itself a digital signal but it may not be recorded in its basic digital form. Although the in and out plugs of a recorder may be labelled as longitudinal time code within the recorder, time code may well be more sophisticated. This is the case with R-Dat. Although appearing to record LTC, it uses the subcodes of the audio data invisibly, producing perfect-looking time code of the type required. In a workstation, time code may not be recorded as a signal but as a reference to cue the replay files. Like most digital audio machines, it will lock primarily to the relevant audio sampling frequency being used; this is modified to time code speed.

Using time code

Time code can be recorded:

1 at the same time as the audio material is recorded;
2 before the audio material is recorded by 'pre-stripping' the tape, and then recording audio and/or video without erasing the time code.

When recording on a video format, time code must start at the beginning of each picture frame to ensure accurate frame editing. This is achieved by referencing the code generator to the video output.

All synchronizing devices need time to lock up but there are particular problems with tape transports. These need 10 seconds of code before programme material starts, pre-roll, to lock successfully. In reality it is sensible to record 30 seconds of pre-roll code.

It is recommended to start a programme's code at 01.00.00.00, i.e. one hour, rather than 00.00.00.00, which avoids the problems of machines crossing midnight time code and becoming confused. In fact, it is wise to start code at 00.58.00.00 58 minutes, allowing time for announcements, tone and colour bars.

Operationally, time code should not be looped through devices by using the time code in and out sockets, for if one fails the rest will; rather it should be divided through a simple audio distribution amplifier and fed individually to each machine.

Recorded time code signals suffer from being copied. They should, therefore, be regenerated rather than just dubbed. Since the code consists merely of 'on' or 'off' pulses, it can be reconstituted, using comparatively inexpensive devices. However, most video recorders automatically regenerate time code, and time code can be transferred between machines without problems. It is important to confirm that longitudinal code is being regenerated by checking the machine's switches and internal menu. Code may stand one direct copy, but this is not desirable. When copying time code using a stand-alone time code generator, both the machines and the regenerator need to be locked to a common sync generator, which could come from the sync output of the video machine; this would apply equally to a digital audio recorder.

A number of methods of regenerating the time code waveform are possible, depending on the severity of damage.

Reshaping time code

To reshape a poor time code waveform, a slice of the wave is first taken through the mid-amplitude point. It is then amplified and clipped, thus re-creating the new waveform. The system is simpler and more reliable than regenerating through reclocking, and is usually found as part of any time code reader (Figure 5.5).

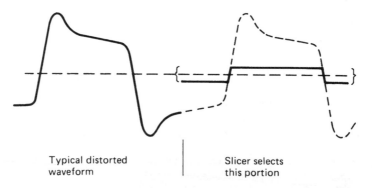

Typical distorted
waveform

Slicer selects
this portion

Figure 5.5 A time code reader in operation. The reader compares the timing between the adjacent edge pairs, the timing being taken where the signal crosses the zero line. The slice is amplified and squared up

Reclocking time code

Time code can be regenerated by reclocking; in this case a synchronous time code generator replaces bad spots in the time code which will not reshape. First, the time code generator is tied up to the faulty time code, then when the incoming code breaks up, the generator code replaces it. The system returns to the original code when it becomes good again, and this code can only be recovered at play speed.

Jam sync

When regenerating a code, the regenerator should be able to reconstitute a missing code if it disappears for a few frames. The device takes the existing code and jams a similar code over it, using the external synchronous input to provide timing information – providing both devices are synched to the same source. Some devices provide an auto mode to provide continuous code should it fail completely.

Advantages of LTC:

- LTC is always available if a tape has been pre-striped.
- LTC can be read at high speeds.
- LTC is suitable for both video and audio recording.
- LTC is a standard system of delivering time code

Disadvantages of LTC:

- LTC cannot be accurately read in stationary mode, or inched and read at slow speeds. To overcome this, time code is often visually 'burnt into' a recorded videotape picture. This is accurate to the frame in stationary mode. (These numbers must, however, be able to be read by eye and then entered into the system via a keyboard.)
- LTC requires a track dedicated specifically to it.
- LTC is subject to severe degradation when copied between machines and has to be reformed. This automatically happens in some equipment (broadcast cassette video recorders, for example).

Vertical interval time code (VITC)

Vertical interval time code (Figure 5.6) is applicable to video recorders. It provides correct time code information when the video machine is stationary and is more useful in audio post-production because of this.

Vertical interval time code is similar to longitudinal time code but without its own separate track. It is inserted into two unseen lines of the video picture. The pulse format which is used doesn't need to be of the bi-phase type, as used in longitudinal time code, since time information already exists within the videotape recorder's own system. The modulation system used is known as a non-return to zero, or NRZ type.

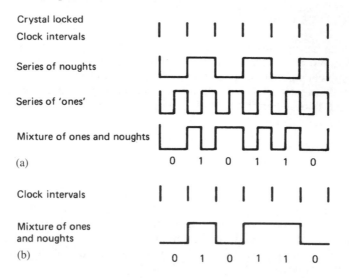

Figure 5.6 (a) The bi-phase mark modulation of longitudinal time code. (b) Vertical interval time code in the television system, which is level responsive; a certain level represents peak white while another level represents black. The black level between the frames is used to represent 0 and a higher level nearer to white represents 1

For practical reasons, the vertical interval time code word has to be recorded twice. This prevents any loss of signal through 'dropout' on the magnetic tape. If minute faults occur in the magnetic oxide in the manufacture of the tape, or through poor physical handling, there is a momentary loss of signal or a dropout. Videotape recorders are able to detect picture dropout and replace a faulty line with part of a previous one, stopping a flash occurring on the screen. This error concealment process is hardly noticeable. It can be even used to replace damaged frames of film transferred to videotape. However imperceptible this system may be to the eye, it is not a technique that can be used on faulty time code. Replacing a time code word with the preceding time code produces obvious inaccuracies. Instead, time code is recorded within the picture frame, more than once, as a safety precaution.

In the SMPTE system, the vertical interval time code signal is inserted not earlier than picture line 10, and not later than line 20. It is also sited in the same positions in both fields. The EBU gives no specific position for VITC words, but recommends that it should not be earlier than line 6 and not later than line 22. International specifications for time code also cover the use of time code in digital video recording (where an eight-bit representation of the code is used, recorded on lines 14 and 277) and within high-definition television systems for the interchange of material.

Vertical internal time code and longitudinal time code are both met in audio post-production using video systems.

Advantages of vertical interval time code:

● VITC is available whenever the picture video is available, even in stationary mode.

- VITC does not require any special amplification; if the VTR can produce a good television picture, it will reproduce VITC.
- VITC does not take up what might be valuable audio track on a video tape recorder.
- VITC is applicable to any video recording system, tape, hard disc, magneto-optical (M-O) drive etc. provided it can record up to 3.5 MHz (i.e. not U-matic low band).

Disadvantages of VITC:

- VITC cannot be read at very high tape-spooling speeds. The individual design of the video recorder affects its ability to read time code at high speed. This is restricted to about 20 times normal running speed.
- To be used successfully in audio post-production, VITC has to be recorded via a transfer from the original videotape.

Midi time code

Midi stands for musical instrument digital interface – a system designed for connecting various electronic instruments together so that one instrument can control one or more of the others. Like the roll on a mechanical pianola, only instructions on what notes (in the form of switching information) to play are given. No recording of the actual sounds is made. It is simply a very sophis ticated switching system with many applications. Midi is well suited to personal computers and, using the correct software, Midi outputs can be sourced through the computer's own printer and modem ports. Being a serial bus system, all the information and data is transmitted on a single cable looped through to the various pieces of equipment. Anything transmitted by a source, or master, goes to all the receivers; each of the receivers is set internally to a specific channel. Very large Midi systems are likely to be run by their own controller. Musicians can use the system to lock together various synthesizers, in the fashion of a multitrack tape recorder, so tracks can be 'recorded' and more sounds built up. Midi is widely used for synchronizing Midi-controlled equipment. The interface is fully digital and works on binary codes, which specify type, destination and the operational speed of the system.

Many digital audio workstations have interfaces which can read EBU/SMPTE longitudinal time code and convert it to Midi. In order to use LTC time code, Midi has to be transformed into data which is compatible. Midi time code has two types of message, one which updates regularly and one which updates for high-speed spooling. In fact, one frame of time code contains too much information for Midi and so it takes two frames to send a frame of information and this then needs to be resolved to frame-to-frame synchronization.

In Midi time code operations, synchronous time information is provided through a sequencer. Sequencers are devices that allow the recording of a sequence of notes or switchings of notes that might make up a song; sequencers need to work on a defined time scale. The sequencer acts as a slave and is fed with Midi time code derived via an interface from the master time code machine (SMPTE/MIDI TC). This can then slave any appropriate Midi device (Figure 5.7).

Most equipment is likely to have Midi in, Midi out and Midi through connections, capable of transmitting information on 16 separate channels. Midi through is used to feed more than two Midi devices, while Midi out is only used at the ends or beginning of a bus. Midi connections, unlike other systems, are standardized by five-pin DIN connectors. The maximum length of Midi cable is about 15 metres. It has become a widely used and successful standard. The Midi interface can be either the master when it generates pseudo time code, or else a slave where Midi commands are derived from the SMPTE/EBU time code readings.

Time code channels

Recommended tracks for longitudinal time code use:

Audio:
  Quarter-inch stereo tape recorder: bottom of track
  Multitrack tape recorder: highest numbered track
  Half-inch four-track tape recorder: track 4
  Two-inch 24-track tape recorder: track 24
  35 mm film: highest numbered track

Video:
  Two-inch quad cue track
  One-inch format, track 3 on some
  U-matic outer track 2
  SVHS track 2

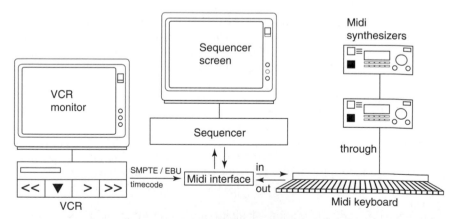

Figure 5.7 A Midi system set-up for music composition using a video recorder with SMPTE/EBU time code. The sequencer synchronizes the time information

Formats with tracks or 'pseudo' tracks solely dedicated to time code:

Audio:
    Quarter-inch centre time code format
    R-Dat
    MiniDisc Cart
    Digital workstations
    Modular digital multitracks
    35 mm outside sprockets (rarely used)
    16 mm outside sprocket (rarely used)

Video:
    U-matic later models: one-inch format
    Some formats: Betacam SP, Panasonic M2, D1, D2, D3, D5, D6, DigiBeta, Beta SX, Digital S, DVPro, DVCam

6

Synchronizers and controllers

Within the terms synchronizers and controllers, there are many devices used in the audio post-production suite – from simple remote controls to intelligent synchronizers designed to control many separate devices. Some equipment, such as audio workstations, includes controllers and synchronizers as part of the system. Essentially, controllers make equipment run together in synchronization to a frame accuracy – that is, to within at least 1/24th of a second. Synchronous dialogue is most susceptible to synchronous problems. From the audio to picture point of view, any synchronous dialogue more than one frame 'out of sync' is unacceptable, although every sound editor/mixer has met the character who speaks 'out of sync' by nature!

Originally in film studio operations it was the frequency of the electricity supply (60 c/sec at 110 volts; 50 c/sec at 220 volts) that was used as a reference to synchronize various pieces of sprocketed equipment together at precisely the same speed. Picture camera and sound 'camera', both recording on sprocketed film, were driven by synchronous motors whose speed was derived from the mains frequency. Once a camera was running at mains frequency and a sound recorder was up to mains frequency, the two would be in perfect synchronization; frame-to-frame synchronous picture and sound recordings could be made. With the introduction of television more accurate synchronization was required and it became common practice to provide a master pulse (frequency) generator to synchronize every piece of relevant equipment. Incredibly accurate, they are driven by a crystal whose oscillations are divided down many thousands of times using the same principles as a digital wristwatch. Devices such as these in smaller, less accurate form control portable camcorders and recorders; in fact, they provide control for all digital recording equipment. Modern audio recording studios will also have their own central generator or use a specific piece of equipment to act as a generator to 'gen-lock' or synchronize equipment together. (This will also provide digital audio signals with their precise synchronized timing to control their data processing.)

Synchronization is achieved when two devices are running together, in what is called lock, perhaps first recording then playing back; this may be time code or sprocket controlled. Although running in lock, the devices may not be running at the same speed. One could be running at 24 frames per second and the other at 25 frames per second, for example. Providing this is always known, the two can be 'synched up' – if the information is lost, it will take some time to unravel the synchronization problem. For audio post-production it is also important that machines are able to be run up to speed from standstill in synchronization as well; controllers and synchronizers offer this function.

The heart of any audio post-production system is the synchronizer and its control unit or human interface. From here, the picture and sound equipment are made to run together locked in synchronization. In fact, any individual piece of audio post-production equipment can be synchronized with another, providing the material has time code recorded on it – SMPTE/EBU or Midi, is pulse coded, or is physically sprocketed. It does not matter how different the synchronizing systems are. If the synchronizing information is available and accurate, synchronization is possible.

Once it was not unusual, in film recording operations, for one company to manufacture and supply all the necessary equipment for a post-production centre providing a well-interfaced running system (this is a tradition that began at the onset of sound motion picture production, with such companies as Western Electric and RCA).

Digital audio workstation manufacturers now provide a similar core facility able to reproduce and record soundtracks with the controller and synchronizer provided as part of the system. Specialist manufacturers who supply the additional equipment such as video players, modular digital multitrack recorders and R-Dat players offer specialist options to allow their equipment to be synchronized in the post-production environments using recognized interfaces and protocols. The functions offered by an audio workstation are shown in Figure 6.1.

Separate standalone synchronizers are, however, also available; these synchronize one or more pieces of equipment to another. Although the cost of these synchronizers has dramatically reduced, it is now often cheaper to purchase equipment with its own interfaces than to buy a separate synchronizer. Unfortunately, it is not possible to cover every eventuality, and many studios need to hire in equipment for a particular production or need to 'lock' their existing equipment to new equipment. Standalone synchronizers are available with every type and combination of interface, from film projectors to music samplers, although many workstation controllers can, with additional software, cover these eventualities. A controller for an audio post-production multiple machine system is shown in Figure 6.2.

The basic function of any controller is to provide motion control and synchronization. This covers single-speed forward, fast forward, fast reverse, and single-speed reverse with additional 'scrubbing' (inching) during picture search, and switching into and out of record.

Essentially, therefore, a synchronizer and controller enable sound and picture transports to run up to playing speed and into synchronization. At both fast and slow speeds synchronization need not be accurate, merely sufficient to allow the devices to stay in step – ready to sync-up at play speed on command. Reverse at single speed and synchronization at standstill are facilities often expected in film operations where equipment runs backwards and forwards in synchronization, for rehearsal and then for a final completed mix. In this mode it is said to be running in 'rock and roll'. To keep tape- and film-

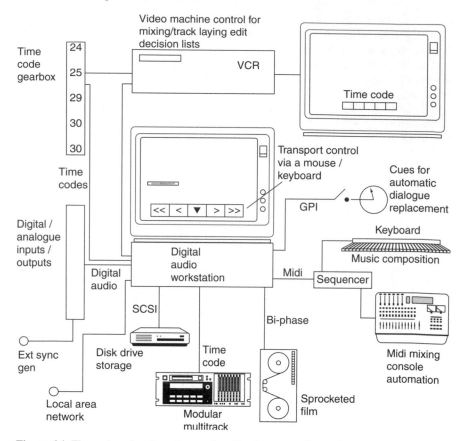

Figure 6.1 The various functions that can be offered by an audio workstation

driven devices in synchronization with an outside source or with their own internal generator, special motor control circuits are used to ensure that the speed of the tape is exactly in step with the controlling source.

Servo control of tape transports

Tape machines (video or audio) and modern film transport drives are synchronized by servo control systems. The drive motors receive their synchronizing information from control pulses or time code recorded on the tape or, alternatively, from a free-running tachometer wheel driven by the sprockets of the magnetic film. These codes or pulses are then referred to an incoming master signal. The machine's speed is then continuously altered, to match the exact speed of the incoming signal. This is called a 'closed loop' servo system, so the precise speed of the transport is slaved to an outside source. In this way, many transports can be run up together in synchronization, with their varying speed and direction controlled from the master.

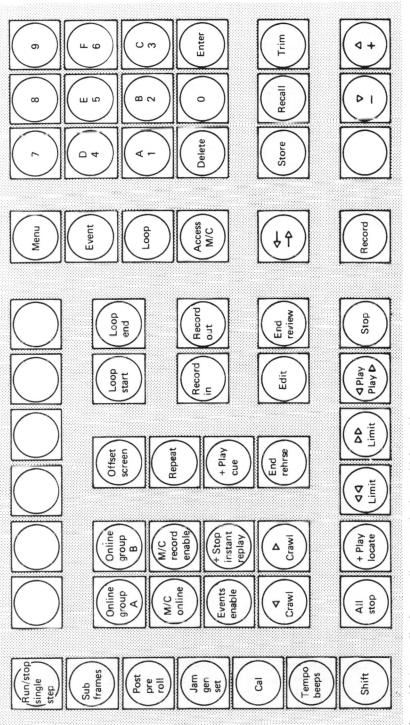

Figure 6.2 A controller for an audio post-production multiple machine system

Master

Various terms are used to describe the machines and the methods of synchronization used for controlling audio post-production systems. At the centre is the master machine. The speed of the master is not controlled by the synchronizer; it produces either time code (perhaps from a video recorder or workstation) or sprocket pulses from a film machine (known as 'bi-phase'), and the rest of the system follows as slaves.

In a simple, but standard, post-production system using only a video recorder and a digital audio workstation locked by time code, the video may well be the master and is operated either manually from its own controls or remotely from the workstation controls. Video tape recorder transports are more difficult to control from external sources and are slow to lock-up as slaves. Workstations and video-disc recorders, on the other hand, can lock-up and follow almost instantaneously makng for more efficient operations in audio post-production. Audio workstations can act as either slave or master.

Slave synchronization

A slave is a machine which is forced to run in synchronization with the master. These can be R-Dat machines, modular digital multitracks, video recorders, or workstations. They chase a given time code or pulse chain. As a slave a workstation, depending on its computer power, will follow variations in lock up to 10 per cent from mean without inducing audio problems. Some are able to operate in reverse in lock, others cannot. The more sophisticated workstations allow remote control of the external master from their own built-in controls. Any system will have inherent problems in maintaining slave synchronization – in an analogue system the speed of recording tape has to be mechanically altered to maintain synchronization, without introducing noticeable speed fluctuations; in a digital system the speed has to be maintained without glitches or clicks. Digital recorders offer far more sophisticated possibilities than their analogue counterparts.

Synchronous resolution defines the limits to which the system will normally synchronize. These parameters can usually be optimized within the synchronizer for the particular equipment used. Some controllers are set in the factory to emulate particular machines, in audio post-production often a videotape recorder. A synchronizing slave will not synchronize exactly – it will tend to lag behind. This lag can range from one frame to sub-frame resolution of 1/80th or 1/100th of a frame or even a sample, depending on the equipment.

Chase synchronizer

Chase synchronizers are found in simple mixing operations, conforming and transferring sound into a workstation, perhaps through a special edit decision list; they are also used for synchronizing a separate sound track with picture for telecine operations.

The simplest synchronizer is the single chase unit either as a standalone unit or as an option to an ordinary machine. These are offered on many machines at a cost of little more than a good microphone. In operation, the machine senses the direction and speed of the time code fed into it and follows or chases the master wherever it goes, modifying the slave's speed to match that of the master. Only time code is needed to control the machine giving speed and direction information, making operations very simple.

Trigger synchronization

In slave synchronization, the slave follows the master up to speed and then follows the master when it reaches synchronous speed. This continuous slaving is unnecessary with much equipment which is capable of running at synchronous speed within its own resolving system. In this situation the slave is triggered to switch over to its own speed control system when the master reaches synchronous speed. This can be unreliable. If a system is only trigger lock it is unlikely, unfortunately, to notice if time code has jumped and has become discontinuous code, due to an edit, because it is no longer actually

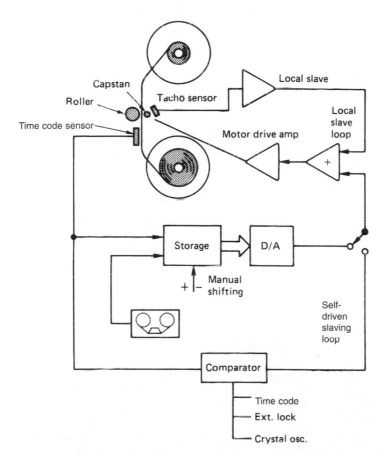

Figure 6.3 A servo control system

reading the code. If, however, the device follows and chases time code, it will detect a time code break, dropping out of play or record or flagging up a warning. A chase and trigger lock will warn of a break but will still freewheel over the problem, usually for up to about 4 seconds; this figure can be set up by the user in some manufacturers' equipment.

Film synchronizers

In film systems the projector is often the master, although a central generator can be used, when the projector operates as merely another slave, with no additional synchronizing problems. The train of sprocket pulses is coded to indicate direction; the pulses vary in speed with the speed of the master, and drive modular multitrack dubbers, audio hard disc recorders or analogue sprocket-drive machines under total transport control. A film synchronizing system is shown in Figure 6.4.

The lock-up or synchronizing time defines the period it takes for the master and slaves to reach synchronization. It should be defined within three parameters:

1 The time synchronous resolution is achieved; this will be determined by the slowest machine – usually the video tape recorder. Video disc recorders

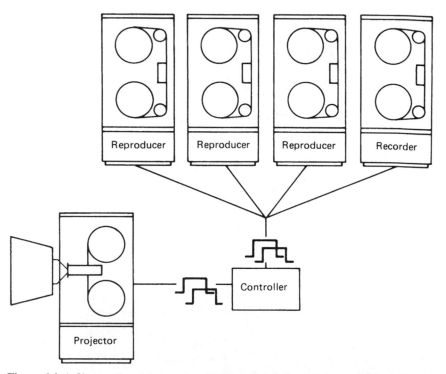

Figure 6.4 A film synchronizing system. Bi-phase signals from the controller/generator are sent to the projector and the sprocketed film transports. The signals provide total transport control of direction and speed

locked to digital sound systems and video recorders integral with audio workstations should have almost instant lock-up, although this may depend on how hard the computer processors are working.

2 The time usable sound is available – in analogue, dependent on reaching specified 'wow' and 'flutter'; in digital, dependent on when acceptable sound is heard (dependent on the sophistication of the processors).

3 The lock time against fast wind speeds, the most crucial operation being the time taken from fast rewind to single-speed forwards. With digital systems this should be almost instantaneous.

Multi-machine synchronizers – E–S bus

This system was the first attempt at a multi-machine control system to international standards. Designed by the EBU, it is a digital remote control system using a common bus. It carries information at 38.4 kbps. It allows any manufacturer's equipment or synchronizer with an E–S bus specification to control any other.

Midi machine control (MMC)

Midi may be used to remotely control audio equipment; it works on similar principles to the E–S bus but is cheaper to implement and as such is an interface found on much semi-professional equipment where it can offer sophisticated facilities. A Midi controller will offer a selection of ports, all of which will sustain a Midi machine control.

Workstations themselves offer Midi in and out connections to allow control over other Midi devices such as musical sequencers, necessary if workstations are to be used for music recording. In operation, Midi input signals are merged with timing information generated internally, and the result is retransmitted at the Midi output socket; usually all types of messages are 'synchronized'. Midi digital control systems often offer a synchronous time code interface within workstations.

Intelligent synchronizers

When only one synchronizing audio machine is required to follow a master video machine or a digital workstation, an integral chase synchronizer may be the ideal device to use. However, when more than one slave is needed, a multi-machine system may be more suitable. This can increase synchronizing efficiency, as well as providing individual remote control for all the machines. Multi-machine synchronizing systems are found as options on more sophisticated audio workstations.

In a control synchronizer all the machines, master and slaves, are connected to the central controller. Information is often incorporated within a machine page in the workstation specifically for configuring the machines to be used. The control synchronizer can be further developed to produce an even more sophisticated system of transport control known as the intelligent synchronizer. These synchronizers are used in traditional linear video editing and memorize the way in which each machine under control responds to commands. The commands are then modified to speed up operations.

The synchronizer may be programmed to follow the ballistics of a particular videotape recorder which slowly reaches speed with the pictures rather than suddenly finding speed; the videotape machines are parked at a set time, namely the pre-roll time, ahead of an address. This allows the machines to run up to speed prior to the address point. This speed will automatically be reduced to the actual run-up time of the slowest machine, thereby speeding up operations. Some synchronizers will chase and trigger-lock machines, running up on frame lock to within 1/100th of a frame, and then automatically switch to their own generator lock.

Uses of synchronizers and controllers in audio post-production

Sophisticated standalone controllers are now only found in analogue multitrack studio or film mixing operations where tracks created in digital workstations are downloaded onto film dubbers or tape for the final mixing of soundtracks.

Synchronizers and controllers found within digital workstations offer a variety of functions. The actual controller itself may have an ergonomically designed, dedicated control panel for ease of operation or just be part of the computer's own interface – a touch screen or a mouse. There may be a selection of macros; that is, a sequence of well-used keystrokes that can be stored together by the operator so that they can be reproduced by using just one or two keys.

Synchronization will be provided for the external video recorder and is usually provided with a Sony nine-pin interface. Other interfaces, such as delivering bi-phase and remote controlling of outboard equipment, may be options. Sometimes it will be necessary to synchronize other devices such as a time coded R-Dat or even a quarter-inch analogue machine playing location tapes. In these situations, it is most likely that the audio would be loaded into the workstation in synchronization and then manipulated within the audio workstation itself, ready to be mixed on the final track.

Controllers, whether standalone or integral to audio workstations, often provide the following features:

- Machine selection. Switching options are available to reassign slaves and masters. This permits any combination of the machines as master/slaves. The control system should also allow any machine, on its own, to be addressed; in addition, systems offer variations in the type and speed of locking.
- Time code control. Some systems offer variations in lock-up times and an adjustable default period before time code regeneration stops after break-up of code. Unstable time code can be regulated to compensate for unstable time code dropout, jitter and suppression of wow and flutter.
- Out of sync alarm. Lights indicate when part of the system is 'out of sync' or time code is corrupted.

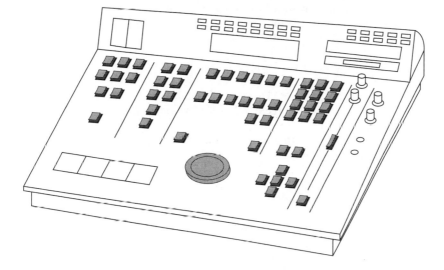

Figure 6.5 A remote controller for a workstation. The workstation is controlled via dedicated switches, not via a mouse, keyboard or touch screen. The large size of the controller is unusual, but effective

- Movable zero. This facility allows the time code counter or footage counter to run from zero at any desired point.
- Cycle or loop facility. The system sets up and repeats a play–rewind–play sequence between points which are held in the memory. It is used in Automatic Dialogue Replacement.
- Trim. This applies small offsets of frames to individual machines or tracks against the master.
- Slip. This occurs when a master and slave are running in synchronization but have been moved apart relative to each other. The facility should allow for an automatic return to synchronization with the offset calculation remembered.
- Events control (GPI). The automatic starting and stopping of external devices such as sound cueing devices, called General Purpose Interfaces.
- Security anti-run-off. Stopping tape-driven reel-to-reel machines running off the ends of reels in high-speed mode.

Problems with time code synchronization with a tape-driven machine running as a slave to a workstation or telecine machine

If a slave machine is running wild when selected, check:

1 that the machine is selected to external drive;
2 that the machine has been run to load the synchronizer with its time code at the correct speed (tachometer-controlled machines);
3 that the time code and audio material are at the same speed;
4 whether a 'go to' position has been loaded into the controller;

5 that there is no unwanted offset in slave memory;
6 that the time code is continuous;
7 that the time code is present.

If time code is not being read correctly, check:

1 cable connections;
2 whether the time code is recopied and of poor quality (second generation may not read);
3 that the time code is at the correct frame rate, and speed;
4 that the time code is at a usable level;
5 that the memory does not include an unwanted offset.

If a slave runs in the wrong direction, check:

1 whether the time code has crossed midnight (00:00:00:00); some synchronizers cannot tell direction across midnight;
2 that discontinuous time code numbers are not being repeated within a roll.

Requirements of digital or analogue video tape recorders suitable for post-production synchronizers

- To successfully interface with the control system.
- To reach play speed quickly and smoothly from any previous mode, so as to assist synchronization with other machines.
- To run at play with as little 'wow' and 'flutter' as possible.
- To run in single-speed reverse.
- To jog picture for search and cueing accurately.
- To deliver usable time code in any mode.
- To play off vertical interval time code if required.
- To automatically unlace the head drum, if stationary after a set time, and so reduce the possibility of tape damage and wear – particularly if picture masters are being used.
- To run at high speed in contact with the head, producing good pictures in all modes and without excessive pixellation if digital.
- To offer remote control of all facilities required.
- To offer excellent audio recording quality if the recorder is to be used as an audio master recorder.
- To play back both NTSC and PAL videotapes.

Some audio workstations provide an integral video recorder of adequate quality as part of the overall system; with these systems the pictures lock immediately to the workstation's audio.

7

Recording audio for post-production

Up to now we have examined the equipment and the electronic control systems used in the audio post-production studio. This chapter examines the processes leading up to the final recording of the soundtrack.

Film and video soundtracks can be reproduced successfully in any type of environment but their accompanying pictures cannot. Generally, film is most suitable for theatrical releases; video, on the other hand, is more suitable for distribution to individual homes, where many small groups of people watch on the smaller screen.

A large projected motion picture image has a high resolution and image brightness that is difficult to reproduce electronically, even using high-definition wide-screen formats.

This means that theatrical films tend to be more precisely made, since technical and artistic finishes are far more easily seen. Indeed, the same applies to wide-screen 16:9 format high-definition television which needs more attention to detail than the standard 525 or 625 3:4 ratio. The production techniques used in the making of motion picture films and electronic television programmes were once very different. Now, with the introduction of computer techniques, the two systems have been integrated at the editing stage, taking advantage of the best of both worlds.

However, from the outset, if audio post-production is to take place, decisions have to be made regarding synchronization systems. For film a simple traditional pulse system may be sufficient; however, time code may offer a more versatile solution to later synchronizing problems. In Europe the speed of time code is a simple choice: 24 or 25. In the USA there are various options depending on the final distribution of the project. Whatever is chosen, the same time code speed must be maintained throughout the project.

Multiple camera videotape recording

Originally, broadcast television pictures could not be recorded or edited, and all productions were transmitted live. Two or more cameras would simultaneously shoot a scene and the various outputs were cut together as the programme was transmitted. The sound was transmitted 'live', as it happened.

In this type of 'live' production, the audio mixer is responsible for the recorded sound balance. He or she mixes together the various sound sources, and may be assisted by an operator who controls prerecorded sounds from perhaps MiniDisc cartridge machines, audio recorders and compact disc players. To assist the sound mixer on the floor, there may be one or more 'boom swingers' operating a microphone on an extending arm. The sound mixer, sitting in the control room, studio or mobile control room, listens to the output of the boom microphone and the other sources and balances them together. Since the boom microphone has to cover all the action as it progresses on the studio floor, the microphone sound pickup needs to be wide, and may sound over-'bright'. This particular problem may be exaggerated with high sets and also bare floors which are necessary to allow cameras to track over them unimpeded. The sound mixer will compensate for this through the mixing console and the facilities it offers.

A studio multi-camera production system is shown in Figure 7.1.

If the sound mixer knows that the programme is to be recorded and then audio post-produced, he or she may decide to leave the process of modifying, and adding additional sounds, until later, well away from the tensions in the control room. Programmes such as light entertainment, situation comedy or a musical concert are often post-produced.

On a large-budget programme, the decision may be made not even to mix together the various video outputs of the cameras at the time of recording but to record each camera's own output separately onto a video recorder of the same format as used in the editing suite (multiple camera isolation recording). Vision mixing can be completed later. Similarly, the output of the various microphones may be recorded onto a multitrack sound recorder in synchronization with the cameras. If only four sound tracks are to be used, recordings can go straight to a video recorder, otherwise a synchronized modular multitrack digital machine would be suitable.

Sometimes, rather than record a light entertainment show in front of an audience with the problems of discontinuous shooting and the actual recording time involved, an audience may be invited to view the final edited show. At this point, reactions will be recorded in synchronization and then mixed onto the final track.

Single camera shooting (video and film)

Single camera productions are shot in a discontinuous manner with one camera shooting all the material needed. The action is recorded; the camera stopped and moved; the lights changed to their best position; then the action continues. In editing, the various shots are then cut together to make a complete sequence. The sound is recorded at the same time as the picture, but it is often incomplete.

Single camera recording more often takes place away from the studio with its acoustically controlled environment. Often sound recordists have only their headphones and experience with which to monitor the quality of the sound, so they try to produce the best possible quality sound from what is, quite frequently, the most unsuitable locations. To produce the 'cleanest' sound they may, if they know audio post-production time is available, record

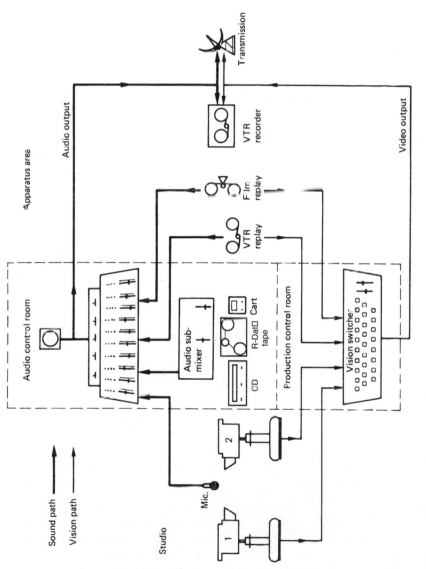

Figure 7.1 A studio multi-camera production system; the audio output can be recorded and post-produced or transmitted live

separately each sound that makes up a scene. Usually the recordist will use a small R-Dat machine or perhaps an analogue Nagra quarter-inch machine; both are available with time code. However, with a quarter-inch Nagra machine, simple Pilotone may be used.

The R-Dat is small and of excellent quality but not as robust as the older and traditional analogue machine. In extreme conditions they may suffer from condensation, and they dislike buffeting and vibrations such as found in ships' engine rooms. The traditional location portable machine was the analogue 'Nagra' which used a pulse synchronizing system similar to video control track; this still finds use in film operations. The machine has been replaced by a larger portable four-track open-reel digital recorder, more suitable for drama rather than documentary use.

For the ultimate robustness on location the solid state recorder is the answer.

An example of location shooting

A simple scene may, for example, consist of two people walking, talking to each other and then driving away in separate cars. The scene will consist of many separate sounds – the feet of the two characters meeting; the talk between the two, consisting of a separate recording of each; the doors slamming and the noise of the two cars starting. In addition, the general background noise of the complete scene is needed. This may be cars going by, birds singing, street vendors or activity in a harbour. Some of these sounds may be recorded in stereo; much will depend on the policy of the production company. Some television companies expect every sound to be recorded in stereo, others such as theatrical film producers expect only the effects and music to be in stereo. The mono dialogue is then moved around the sound field as required in the final mix. The sound scene is more than just one picture following another; it consists of sounds that geographically, and in time, knit together the whole discontinuous scene. The amount of post-production time available will influence the sound recordist in his or her approach to the job. The more time available for audio post-production, the more additional material can be recorded, and the 'cleaner' the various recordings will be. It may even be an advantage to record more than one track of sound during a take.

Single camera recording with multitrack separate sound

There are certain advantages of recording sound on more than just one track of the recorder, whether in a studio or on location. If a scene consists of various sounds all happening at the same time, these can each be recorded separately onto the tracks of a digital multitrack audio recorder, and later mixed more accurately in a studio away from the tensions of the set. On-screen lines can be recorded on one track and off-screen lines on another, or, if the artistes are fitted with their own personal miniature radio microphones, each can be recorded onto a separate track of a synchronized modular eight-track digital multitrack machine, as the action takes place. This suits ad lib scenes where the director wants to record many people talking at once for later mixing down at the appropriate levels required. Directors have been known to record up to eight separate voices in these situations. This, however, creates immense problems with interference between the various radio microphone channels on the location. All this allows considerable flexibility in post-production, although much depends on the time available.

Even a two-track recorder can be used to advantage, particularly when two artistes are speaking with very different microphone sound quality. For example, one may be recorded with a personal microphone under many layers of clothing, producing a muffled sound, the other with a personal microphone which is hardly hidden. These two very different sounds can be recorded on separate tracks of the tape recorder, and later 'sweetened' in audio post-production, where the sound quality can be judged critically. Similarly, action which can only be recorded once, such as an explosion, can be recorded in close-up on one track, and in a long shot on another.

Music is often recorded on location using a multitrack audio recorder. Here, individual instruments, or groups of instruments, can be recorded onto the various tracks of the recorder for mixing later, in a music studio. Normally, a rough mix from various tracks is provided at the time of the recording. This is transferred onto a synchronous R-Dat for loading into the editing workstation for use as a 'guide track'. The original tapes might be brought into the audio post-production suite for final mixing, or alternatively, while editing takes place, the music is mixed in a specialized music recording studio.

Identification and slating

When using film cameras, the sound and picture are recorded separately. This is called double system shooting. To match both of these together for editing later, it is normal to provide a synchronous mark on the soundtrack and on the picture. (The two machines only run together in synchronization, at speed.) The accepted way of doing this is by using a clapperboard. Before the director starts the action, he or she asks the sound recordist to 'turn over sound' and then the cameraman to 'turn the camera'. When both the camera and the recorder are running at the synchronous speed, the director calls 'mark'; at this point, a board, on which is written the scene and take numbers, is shot visually, and then a hinged clapper is closed on the top, making an audible signal. Later, these two can be conformed into synchronization or 'sunk up'. The numbers on the board provide information on the slate (the shot number) and the take (the number of times it has been filmed). The board is also quite likely to display time code; this will be either from a central generator or from the audio recorder, and gives a reference of the visual picture position to the audio tape which is also recording the code. The same code may also be recorded on the camera film for later reference. Some systems will also record identification information giving the roll and take number in the time code user bits.

Sound for film

Traditionally, film shot in the field was recorded on a 'Nagra' recorder with synchronization controlled by Pilotone or Neopilot pulses. These synchronization systems send a train of pulses from the film camera to the separate sound tape. Each pulse represents a sprocket of the camera film.

Originally, the signal was derived from a generator attached to the camera motor and sent by wire to the tape recorder. Various methods were developed allowing the camera and recorder to operate independently. The generated signal can be radioed from camera to recorder. Alternatively, the camera can be driven from a crystal oscillator which runs at the same speed as a crystal oscillator at the audio recorder – the latter producing pulses in synchronization with the camera drive. The system provides a simple, effective method of synchronizing film and sound. However, this electronic sprocket recording provides no time information, only synchronizing information.

A quarter-inch to sprocketed film synchronizing system is shown in Figure 7.2.

Nowadays, most film is shot and then transferred into video computer systems for editing. The audio tapes and the film can be automatically synchronized (autoconformed) at the time of the transfer of the film to video, using time code recorded on the film with the same time code on the audio tape. This, however, can cause damage to the film negative during transfer as the telecine machine stops to wait for the audio to synchronize. This method only works successfully if the location sound tapes consist only of sound appropriate to the picture and there was a sufficient run-up time of at least 10 seconds to allow the system to reach speed. If synchronization does not take place in time, the system repeats the same autoconform with danger of damage to the camera negative. This method places constraints on the crew who have to use long run-up times (pre-roll) after turning on the camera and before shooting. The sound recordist must also use two recording tapes, one for synchronous sound and one for non-synchronous sound. A better practice is to synchronize sound after the picture has been transferred. This places no unusual constraints on the crew. To make audio 'synching' simple, audio time code is recorded in record run; this gives the tape a continuous time code signal despite the stopping and starting of the audio recorder.

In the video production, the sound will be recorded onto the video recorder carrying the pictures where usually four tracks are available (depending on format). It is unnecessary to use the clapper here; but, of course, the slate and take numbers are still relevant, providing quick visual information with the time code referred to them. Separately recorded sound effects and dialogue are also identified as individual slates.

Identification and logging

Correctly identifying and logging all the material recorded on location is crucial if subsequent picture and sound editing is to proceed smoothly. Reduced shooting times and fewer production personnel on location (a production assistant may not be employed) make accurate logging sometimes difficult and, with insufficient information, post-production slows up. Time spent on correctly identifying material will save on post-production time. Some camcorders allow good and bad take information to be stored on a special data track on the tape or in memory within the cassette. Systems have been developed that electronically log shots and sound on a laptop computer

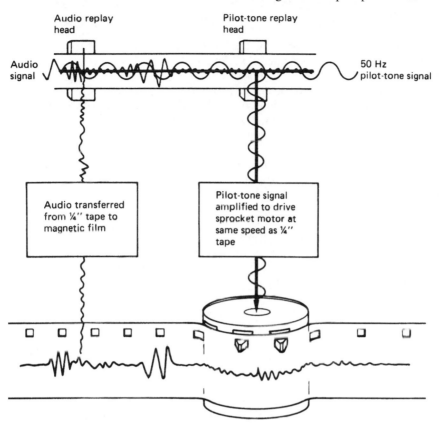

Figure 7.2 A quarter-inch to sprocketed film synchronizing system. The 50 Hz synchronizing signals recorded on location tapes are picked up at a special replay head. The pilot-tone signal is recorded in opposition to the audio system and is 'silent' to it

fed with radio-linked time code from the camera – eliminating the familiar 'What's the time code, please' call from the production assistant. From this an electronic editing list can be produced which can be read by an edit controller. However, systems like this are not a replacement for good planning and shooting practice; they are essentially an aid to improve efficiency. All sound recorded on location must be well identified, and follow the local practice. What tones should be recorded on the head of the tape? Is there any point where clipping could take place? Should the stereo be in M & S or AB format? If R-Dat is used, what is the correct sampling and quantization rate? Can all the material be recorded on 'one generic' tape, or should non-synchronous sounds be separate? Is a specified run-up time required before a shot starts, to allow for later copying to another audio format? What time code rate is being used?

A note should be made if there are close-up sounds, long shots, rehearsals, wild tracks, guide tracks (particularly not for inclusion in the final production) etc. With time code systems, the time can also be logged as the sound is recorded.

Labelling should:

- be legible and according to the local practice;
- give full details of the production, date, personnel;
- give the reel number;
- give technical specifications of format and track allocations;
- identify the recording machine (in case a technical fault is discovered later) giving time code rate and bit and sampling rate;
- give details of line-up recording tones;
- indicate stereo format;
- indicate, if appropriate, which takes are to be transferred for editing (circled takes), which are wild tracks, tracks for fitting, etc.;
- carry identical marks on both container and recording.

Double system or single system?

Modern video recorders record digital and analogue sound to a very high standard, and it is only necessary to record sound on a separate machine if there are insufficient tracks on the camera format – which usually offers four. In film production, however, a separate sound recorder is the norm (combined magnetic sound film recording is unacceptable, both from an operational and a quality point of view). Recording sound with a sound recorder independent of the camera, using crystal control on both devices, with no lead between the cameraman and recordist, has certain operational advantages.

- In certain situations it is physically safer to record sound without an interconnecting lead – such as on small boats or on scaffolding, where extra cabling may prove dangerous.
- Without lead restrictions, the camera, recorder and microphones can be easily placed wherever required.
- Complete sound recordings of current affairs events are possible without breaks. Film cameras and video cameras used in these situations often have a limited recording time (typically 12 minutes for film and 30 minutes for video).
- The sound recordist can record sound whenever he or she wishes, without needing access to turn on the camera; these sounds may be wild tracks, sound-only interviews, etc.
- With multi-camera video shoots using camcorders, it is often difficult for the sound recordist to feed the audio signal separately to two or more roving cameras; separate sound here provides a master soundtrack.

Multi-camera film shooting

Film cameras can be used in multi-camera set-ups, similar to video set-ups in, for example, the recording of music concerts. No attempt is made to produce an edited master, as might usually happen in a video studio. While shooting, each camera may be run continuously or as required. For these purposes, special time code systems have been developed. These allow any number of cameras to be started or stopped during a shoot, without the traditional synchronizing clapperboards being used. During the telecine transfer of the picture, picture time code is recorded onto the video recorder and later this is synchronized automatically, with the audio, since the sound and video picture have their original time codes as recorded on location. In similar situations, time code can be used equally successfully but in a rather less sophisticated way. Each camera is crystal quartz controlled to a precise speed. The start of each shot is identified with a time code figure, recorded by photographing a jumbo-size, displayed time code reader. This is fed from the output of the crystal-controlled master generator which also synchronizes the audio recorder. The display is centrally placed so that all the cameras can see it. Each time a camera is run, the display is filmed and this record of the time code provides a reference for the editor in 'synching up'. The editor matches the picture to the sound which has been transferred with time code.

Time code generators and cameras

Where a time code generator is fitted as extra 'outboard' equipment to an existing video recorder, it is important to ensure that the time code is synchronized to the video pictures, so that each code word begins at a specific point in the field blanking interval, otherwise the code may be unusable in editing. When using time code, other information can be added to the coding through the 'user bits', for example the reel and scene numbers.

Time code generators used in conjunction with camera equipment on location can be set in two separate ways, record run and free run, and both can affect audio post-production operations.

Record run

In record run, the time generator is set at the beginning of the day, or alternatively at the beginning of each tape, and can be used as an accurate timer of tape length. The code generator only runs when the audio tape machine is recording, thus the time code recorded on the tape is, when played back, both continuous and consecutive. This can be important if an audio post-production synchronizer is later used. Time code can only be successful in synchronizing if there is a period of uninterrupted code prior to a take for running up into synchronization. Record run gives this pre-roll time even though the early transferred audio may be from the end of a previous take! Only at the start of the tape is it then necessary to have any run-up footage.

Free run

Time code in free run mode produces discontinuous code. In free run mode the time code generator is set to the real time of day using an accurate clock. It is started and left to run like a clock, continuously. Other equipment, such as a second camera or an audio recorder, can be synchronized to the same time code. However, it is possible that none of these time code generators will have sufficient accuracy to hold synchronization over many hours. There can be problems, particularly with some video recorders, if, for example, the heater is switched on or the batteries are changed, and it is advisable to carry out an accuracy test prior to shooting to check the reliability of equipment.

Synching-up material with inaccurate time code can cause immense problems in audio post-production. Alternatively, time code can be radioed to the various recording devices from a central generator.

High-quality time code generators can be accurate to a frame over a day.

In free run the time code generator acts as an accurate stopwatch, timing events as they happen, irrespective of whether the recorders are on or off. This is useful for current affairs programming, such as conferences or sport, where it is important to know when selected events happened. Since the time code is tied to the real time of day, notes can be made from a digital wristwatch without access to the camera. Using time of day allows everyone to make their own time code notes, without physically examining the camera's time code position.

Independent locked running

Video and analogue recorders are often run independently but controlled by their own crystal-controlled generators. With quarter-inch analogue machines this is satisfactory; however, R-Dat machines are, by design, not particularly stable and, although accurate enough for short takes, domestic and semi-professional machines should not be used for takes of a few minutes or more. It is very important to test time code accuracy if running independently for more than a few minutes.

Recording dialogue at time of shooting

It is the responsibility of the sound mixer to record the best possible sound, even under the most difficult conditions. Sound effects can be re-created in audio post-production, but dialogue is far more costly and difficult to re-record satisfactorily. Therefore, the sound recordist should primarily concentrate on producing high-quality dialogue recordings.

The well-recorded dialogue track will consist of only those sounds that the sound mixer wishes to record. He or she will decide on the 'fullness' of the tracks by taking into account the audio post-production time and facilities available. The ideal dialogue track for audio post-production will:

- be recorded without reverberation. Reverberation cannot be taken out of a recording but it can easily be added. (In stereo, reverberation may be part of the sound field.)
- be recorded with only one artiste speaking at a time. Speech should only overlap for good reasons – such as an argument.
- be consistent in level, quality and stereo position between takes. This means fewer changes have to be made in audio post-production, which can be time consuming.
- not be recorded too wide if stereo is used.
- have sufficient sound at the beginning and end of the dialogue to allow 'loose' cutting.

If it is impossible to record dialogue successfully because of poor acoustics or high background noise, the sound recordist may decide to record the dialogue again. This will usually be done after the original take, as a track for 'fitting', without the camera running, using a more controlled location or just 'closer' sound. The editor will then attempt to fit this against the original dialogue. It is obviously important that the actors record the dialogue with the same intonation and at the same speed. They must also use the same words; often actors unwittingly use different words for different takes, and this can make 'fitting' particularly difficult!

Once the dialogue has been shot, the numerous other sounds needed for audio post-production will be recorded, such as spot effects and wild tracks.

Spot effects

Spot effects are often added in audio post-production from the location tapes. Keys being turned in a car door, for example, might be recorded separately as a spot effect on location, and added during post-production rather than being recorded as part of a dialogue sequence. While dialogue is being spoken, the key effect may be of a very low level and masked, for example, by the background noise of cars going by. If added in post-production, the key effect can be increased in volume to the level required. Spot effects benefit from being recorded separately from the main track.

Wild tracks

Wild tracks, or atmospheric recordings, are background recordings to scenes. They provide the atmosphere of a location and are 'wild' or non-synchronous. Wild tracks are also used to help cover the problems of variations in background level between edited shots. When dialogue is recorded discontinuously, the background level between shots is likely to vary. The dialogue levels will be the same, but as the picture cuts the background sound levels may change suddenly and unacceptably. An 'atmos' track can be added to mask this 'stepping' effect if it is played at the level of the shot with the highest background noise.

Recording effects separately on location is particularly important when moving objects which produce their own sounds, such as cars, are used in productions. Very often, synchronous recordings of cars are incomplete and so extra wild tracks have to be recorded. These wild tracks should consist, for example, of various shots of the car coming in and out of vision, the door opening and closing, the engine starting, ticking over, revving, stopping, pulling away and changing gear, together with interior shots of it running.

Shooting to playback

This is a technique used in musical productions, where discontinuous single camera shooting is used. A musical number will often consist of various shots at different angles, allowing the camera a great deal of freedom. To cover these shots successfully with microphones and record quality sound is almost impossible, since the microphones would have to be positioned for the widest shot used. So rather than record 'live', the sound is recorded in a studio under controlled conditions, or even at the location, using the best possible microphone positions. It is then played back to the artistes, who mime while the camera roams unhindered, stopping and starting whenever the director requires. In a well-organized shoot, the director will know precisely how to shoot each section of music. Tapes can then be prepared against a shooting script. In film making, to assist editing, three beeps of sound are recorded on the tape prior to the particular audio take. This allows the camera assistant to mime the last beep, with a clapperboard in front of the camera. In this situation the playback tape recorder and the video or film camera are synchronized together, possibly locked to one central, external generator, or each running under its own crystal oscillator.

Other methods can be used:

1 A playback tape recorder sends audio simultaneously to the location or set and into a second recorder which re-records playback sound synchronized to the camera. The material is edited in the normal way, using the re-recorded sound.
2 The playback audio machine runs in synchronization with the camera, using time code. This time code is also fed to an audio track of the video recorder. Using separate sound editing, the pictures are then automatically locked up to their appropriate, separate playback tape through the time code.

Whatever system is used, identification and filing have to be carefully considered. When the shooting is completed, the original music tape may well be re-mixed during the final re-recording session, and synchronization must be maintained.

8

Video editing picture and sound

To understand the processes of audio post-production, it is necessary to be aware of not only how material is recorded on location, but also how the pictures are edited. This is closely tied up with sound editing.

Once upon a time, film and video production traced two very separate paths, from location through editing to transmission or release. This has changed dramatically. Now, after the initial location shooting, film and video productions often follow the same editing path using computer-controlled digital hard disc editing systems. Here the material is loaded into the machine and edited, with the advantages of instant access, without having to spool through tapes or film to find shots. This is called non-linear editing. (Some cameras can record directly onto computer discs, instead of tape, saving loading time.) Linear editing is, however, the traditional way of editing video tapes; it is used in quick turnaround situations, for some news operations, for the final editing of multi-camera shoots which are already vision mixed, for producing instant graphics, for some corporate productions, and in education – essentially for fast and often simple editing projects. The high cost of broadcast quality hard disc computer editing systems also encourages the use of simple linear editing systems.

In news operations, where speed is essential, traditional editing suites allow the instant use of video tapes straight from the camera. Editing might well be on a laptop editor at the location, for instant transmission via satellite. The laptop editor, like any video editing suite, in its simplest form consists of two video machines, one a player and the other a picture recorder, with a controller. As suites become more complicated, other equipment, such as a caption generator, another video player and a vision mixer, can be added, providing facilities to produce a complete programme. In video editing, the picture and sound are copied one after another in a linear manner.

A simple editing suite is shown in Figure 8.1.

Originally, video editing systems used the picture synchronizing track of the videotape recorder to control the editing process. This track is in the form of electronic pulses, recorded along the edge of the tape. In a similar way to film sprockets, these can be set to zero on a time counter. They can then be used to control the video editing process. However, these pulses cannot be

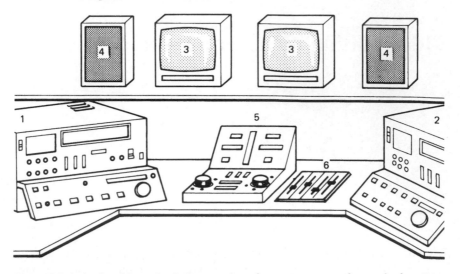

Figure 8.1 A simple editing suite. 1, Cassette player for camera masters; 2, recorder for editing onto; 3, monitors to view pictures on; 4, loudspeakers; 5, edit controller; 6, audio mixer for controlling soundtracks

read accurately to the frame and were superseded by time code which provides frame-to-frame editing accuracy.

Videotape editing is carried out by copying pictures from a replay machine onto a recorder. Two specific methods are used, assembly editing and insert editing. In the assembly edit mode, pictures are added to already existing recordings; each picture is complete with its own time code (recorded at the shoot) and the synchronizing control pulses needed to produce steady pictures. However, when another picture is added to a previous one, the control track pulses are now discontinuous. Each time the picture cuts, the synchronous control information changes. This is likely to produce unstable pictures at the cuts.

The diagrammatic version of a simple videotape editing suite using time code is shown in Figure 8.2.

In the insert mode the control pulses are continuous, so there is no disturbance of the edit. However, to achieve this, the tape has to be prepared for an editing session by pre-striping with continuous control track and time code (Figure 8.3). (This method is used in certain specialist requirements for digital audio tape formats.) The pictures recorded onto the tape are locked to the pre-striped track, and there are no frame rolls on cuts. The time code must be locked to the control track to ensure that editing takes place in synchronization with the control pulse. However, the time code from the original camera masters does not appear on the edited videotape, unless special provisions are made (such as inserting the numbers visually into the picture or recording it on an audio track).

Insert editing is used in both analogue and digital video editing systems; however, digital recorders rely heavily on processing to produce pictures and processing can take time. Simple compact laptop editors may not be powerful enough to respond immediately to commands and frame-accurate editing may not be possible.

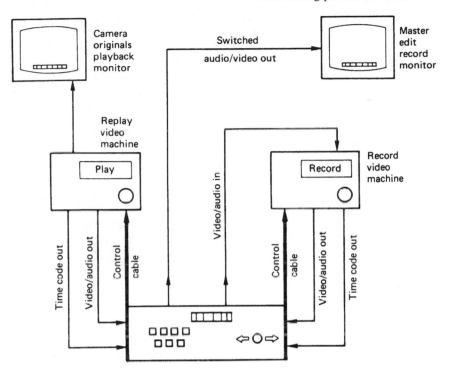

Figure 8.2 The editing system for a simple videotape editing suite using time code

A video edit progresses as the picture is copied from the playback machine onto the recording machine. The machines are controlled from an edit controller which may include a computer keyboard and a video display screen. Shots are chosen visually, but they can be found by giving the player known time code location addresses. Having selected the edit, the necessary time code readings are entered and the edit computer then rehearses the sequence. Once the completed edit rehearsal, or preview, has been performed success-

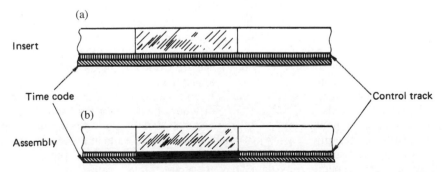

Figure 8.3 (a) Insert editing uses a pre-striped tape with control track and possibly time code already recorded on it. This track remains undisturbed during editing. (b) Assembly editing adds a new control track with each picture edit

fully, the system goes into record. The next edit is added to the previous one, and so a sequence is built up one shot after another. As necessary, special video effects may be added, perhaps running in two or three video machines together. This may mean that a video delay is introduced automatically through the system to eliminate the possibility of frame roll as one video signal is run with another. A result of this can be a loss of audio and video synchronization (similar problems exist when pictures are sent by satellite and sound by land-lines). To restore this, an audio delay line may be necessary in audio post-production. The edit will continue, shot being added after shot.

This is a linear edit – if a decision is made to change the length of a previous shot, the whole sequence following will have to be remade. This is a time-consuming process, possible only if one can remember which source tapes the shots came from. To eliminate this problem, some edit controllers monitor and remember all the actual edit points. They can then produce an edit decision list (EDL), automatically putting this information on paper or on a floppy disc.

Edit decision lists (EDL)

Edit decision lists memorize the way an edit is put together. Every shot is memorized via its original time code readings – the in and the out points. This information allows edited material to be remade at a later date exactly as it was edited. To reassemble an edit correctly, an EDL must be efficient and 100 per cent accurate; in traditional video editing this can be a problem, so special programs have been produced to 'clean up' EDLs. The first cleaning program was called 409. When using non-linear editing systems, this is unnecessary as the EDLs are always clean. To conform an edit successfully, the edit controller must switch the edit master into and out of record at precisely the right frame, whether it be picture or soundtracks. Since the final edit is likely to have been made up of more than one tape, these must also be identified. The same time codes may be repeated on the various tapes used. Reel information can be gathered automatically if it has been identified within the time code user bits on the tape and the controller can decode it. In some formats, reel information can be held in the cassette itself. Alternatively, the controller may need the reel number to be entered manually via the keyboard. Other information can also be entered manually, such as the position of other material needed for the edit, where wild tracks are, dialogue for fitting and any useful sounds for the sound edit. However, some EDL systems do not accept written information and may crash if they receive it! In workstations the information is entered as the picture and sound are transferred in.

The ideal method of re-creating the EDL edit would be for each reel of tape in the edit to have its own player; the edit would then go forward with each player finding the picture and sound as needed and playing it onto the master edit, in almost real time. This is just not practical so the edits are usually built up from one or two machines. There are various methods used to record the video and sound specified by the EDLs; these are called modes.

In mode A, the EDL starts at the top with the first source cassette on the machine, performs the edit and then calls for the next cassette to be loaded for the next edit; this is done manually. It performs the edit and then stops again, demanding the next reel. In mode B, or checkerboard assembly, a cassette is loaded and all the appropriate shots are recorded off from that tape – leaving spaces for other shots from other source cassettes. The next reel is loaded and fills in more spaces with the appropriate audio – and so on.

When loading up an audio workstation either method can be used. The audio workstation picking up sound from the source video or audio tapes with time code can operate as a mode A or mode B controller of the video or audio player. Alternatively, all the material can be fed into the workstation and it can, with the correct software, internally produce the correct edit from an EDL.

Edit decision lists are the principal way of documenting time code generated during editing. In operation, the shot to be used is chosen and the video and audio tracks selected, time code is automatically logged.

The source tape providing the first shot is used and documented automatically as:

IN   04:12:00:00 V (video record) A1 (audio track one record)

The out point is then chosen; the edit is logged automatically as:
OUT   04:12:15:04 V A1

This is the source out.

The length of the shot is:
Duration 15:04 secs V A1

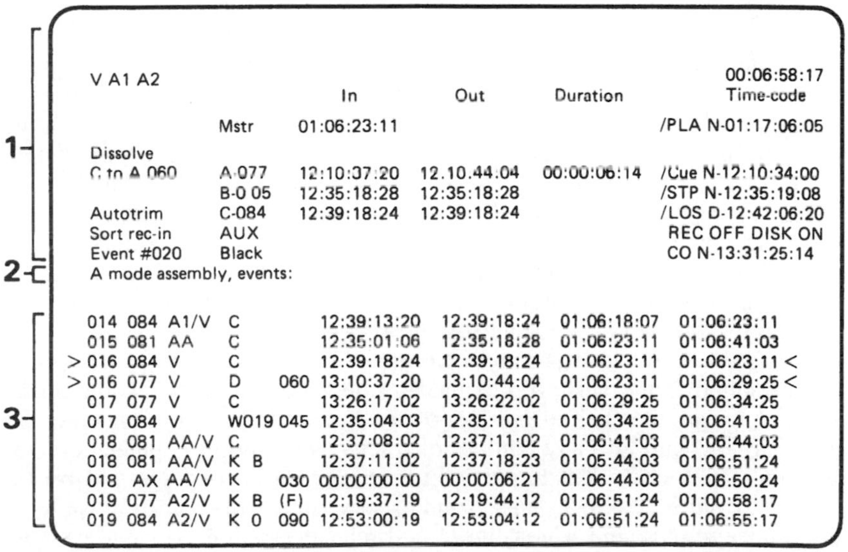

Figure 8.4 A CMX edit decision list. 1, Menu area; 2, system message area; 3, edit decision list

On the master edit created from the source material the:
  Record in is documented as 01:00:00:00 V A1 (programme start)
  Record out is 01:00:15:04 V A1 – edit continues

The computer will store other information:
  edit number
  reel number
  edit mode Video recording V1
  Audio 1 track recording A1 Audio 2 track recording A2
  Audio 3 track recording A3
  Audio 4 track recording A4
  Plus additional information

Each edit is noted with a number.

The information is recorded automatically by computer and produces a printout in the form of an edit decision list usually in a standard format such as CMX, Sony or Grass Valley. Figure 8.5 shows an edit decision list, audio only.

Videotape formats have available up to four tracks of sound, and these can be used to good effect by the video editor as the picture edit progresses. Unfortunately though, in some older formats (Beta SP and M11), two of these tracks are only audio frequency modulation tracks, and have limited use. These tracks can only be recorded while picture is being laid down, thus sound cannot be extended into the 'next' picture, nor recorded as a sound-only track – without, of course, wiping the already recorded picture! They can only be heard in play single speed mode. Early industrial formats have only two independent audio tracks available.

Bouncing

Within the video editing suite it is possible, with good planning, to mix simple soundtracks in mono, with satisfactory results. Using a two-track video recorder, the edited sound is recorded onto track 1, with the levels carefully set not to change violently. When the edit is completed, further sounds can be added. The edited videotape's audio channel 2 is selected to record, and through a sound console the already recorded audio from channel 1 is 'bounced' onto channel 2 with other audio such as narration and music to complete the mix. This can be extended further by making a copy of the final edited master onto another videotape (with time code) and then recording separate soundtracks on each of its two tracks. The master and the copy videotape are then played back together in synchroniza-tion. The sound from track 1 of the master videotape and the two soundtracks from the copy are then mixed together on track 2 of the master. In this method it is possible to build up any number of tracks using multiple tapes; alterna-tively, the tapes could be dumped onto a multitrack recorder and mixed in a sound suite. Laying many soundtracks in a video editing suite is usually uneconomic and a non-linear editing workstation is a more viable proposition.

```
TITLE:    COLOURS AUDIO ONLY
001    101    AA     C      01:07:03:10 01:08:05:10 10:00:00:00 10:01:02:00
*FROM CLIP NAME:    WT WAVES CLOSE M&S:
002    DAT    NONE   C      15:51:36:00 15:52:32:22 10:00:05:01 10:01:01:23
AUD    3      4
*FROM CLIP NAME:    TITLE TRACK
003    DAT    NONE   C      15:53:17:04 15:53:35:20 10:01:01:23 10:01:20:14
AUD    3      4
*FROM CLIP NAME:    TITLE TRACK
004    DAT    NONE   C      16:10.07:00 16:18:27:19 10:02:19:01 10:02:39:20
AUD    3      4
*FROM CLIP NAME:    GUITAR LINK
005    007    AA     C      07:10:43:03 07:11:22:02 10:02:23:10 10:03:02:09
*FROM CLIP NAME:    WT-RAIN/GALLERY/WINDOW*
006    DAT    NONE   C      15:56:47:07 15:58:17:07 10:04:04:18 10:05:34:18
AUD    3      4
*FROM CLIP NAME:    KISS/MORNING
007    101    AA     C      01:10:43:23 01:11:46:01 10:06:19:11 10:07:21:14
*FROM CLIP MAME:    WT WAVES DIST M&S*
008    DAT    NONE   C      16:15:06:01 16:16:52:10 10:06:19:11 10:08:05:20
*FROM CLIP NAME:    TEAUCH SAID CHOOCH
009    004    AA     C      04:10:13:13 04:10:14:23 10:07:26:12 10:07:27:22
8FROM CLIP NAME:    7A-FWT*
010    DAT    NONE   C      16:10:13:22 16:11:01:07 10:08:57:06 10:09:44:16
AUD    3      4
*FROM CLIP NAME;    PAINT
011    101    AA     C      01:08:37:10 01:08:48:19 10:11:20:00 10:11:31:09
*FROM CLIP NAME:    WT WAVES CLOSE M&S*
012    DAT    NONE   C      16:04:07:05 16:05:03:16 10:11:20:00 10:12:16:11
AUD    3      4
*FROM CLIP NAME:    DARK
013    101    AA     C      01:07:32:20 01:07:43:01 10:11:42:15 10:11:52:21
*FROM CLIP NAME:    WT WAVES CLOSE M&S
014    006    AA     C      06:23:40:21 06:23:46:08 10:11:57:15 10:12:03:02
*FROM CLIP NAME:    9A-WT*
015    DAT    NONE   C      16:05:02:16 16:05:24:00 10:12:49:05 10:13.10.14
AUD    3      4
*FROM CLIP NAME:    DARK
016    DAT    NONE   C      16:03:44:16 16:04:19:18 10:13:14:15 10:13:49:17
AUD    3      4
*FROM CLIP NAME:    DARK
017    DAT    NONE   C      16:18:10:01 16:19:14:01 10:16:22:12 10:16:46:12
AUD    3      4
*FROM CLIP NAME:    GUITAR LINK
018    DAT    NONE   C      16:06:49:02 16:07:08:00   10:16:55:13 10:1714:11
AUD    3      4
*FROM CLIP NAME:    SCAT
019    DAT    NONE   C      16:07:08:01 16:07:14:01 10:17:24:17 10:17:30:17
AUD    3      4
*FROM CLIP NAME:    SCAT
020    DAT    NONE   C      15:54:23:18 15:55:01:17 10:17:30:17 10:18:08:16
AUD    3      4
*FROM CLIP NAME:    CREDITS TRACK
021    DAT    NONE   C      15:56:04:21 15:56:10:07 10:18:08:16 10:18:14:02
AUD    3      4
*FROM CLIP NAME:    CREDITS TRACK
```

Figure 8.5 An edit decision list, audio only *(Courtesy of Waterfront Studios, Glasgow)*

Non-linear picture editing

Traditional linear video editing has many basic problems:

- The tape cannot be cut so it is impossible to simply shorten or lengthen a project.
- Every insert must replace a sequence of a specific length.
- Tapes need to be copied as part of the process, which is time consuming and leads to loss of quality.

Non-linear computer editing eliminates these problems and has virtually replaced traditional complex video and film editing. Non-linear editing allows an editor to choose the pictures required and edit them entirely within the domain of the computer, replacing pictures at any place in the edit and at any length; every component of the edit is available instantly. These are the great advantages of digital random access hard disc editing. The disadvantage is the expense, and in films for theatrical release, the inability to see the edit on any but a small television monitor.

Non-linear editing systems that can record broadcast quality pictures and sound with sufficient space to carry all the material shot may cost as much as a traditional video tape editing suite, but there is increased versatility. Originally only designed to replace simple editing facilities, non-linear editing systems now offer graphics, digital video effects in multiple layers and four or more audio tracks – in fact, all that is necessary to produce a complete production. These facilities will become more and more sophisticated. A simple non-linear picture editing workstation is shown in Figure 8.6.

Although edits are created in non-linear editing, they are never physically 'made' – all the material remains in the system unaltered. Sound and picture

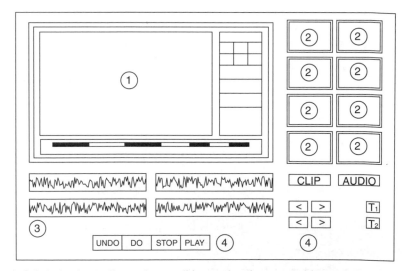

Figure 8.6 A simple non-linear picture editing workstation operated by a mouse. 1, Editing screen; 2, clips for preview; 3, audio tracks; 4, motion control

edits consist only of an edit decision list which the computer updates each time a new edit is created.

Non-linear picture editing systems always include sound editing facilities, but of varying sophistication. They are operated with two or more screens. In simple systems, one screen may show the edit page, the selection of shots available, the transport controls, and a graphical representation of the pictures and soundtracks. The second screen is an ordinary television monitor showing the edited picture. More sophisticated systems divide the various operations into three or even more screens. Non-linear editing systems vary in their facilities and complexity from one manufacturer to another. They follow a general operational pattern.

As with any editing system, it is essential, if the process is to work efficiently, that all the material is logged. This may be only a handwritten log, or it may be a computer-generated log. With the correct software this can control the operation of a non-linear editing system. This log will identify all the material needed for the edit. Material not required – NG takes – will be omitted and only those needed will be transferred into the workstation, reducing the amount of video disc recording space used. The log will contain the master tape's unique numbers, time code in and out points (omitting discarded takes) and written information about the takes that are required in the edit. Transfer is automatic, controlled by time code. As each time code point is reached, the workstation records the picture and sound as a take or clip, noting its time code identification and any written information. The first frame of each of these takes or clips is shown on the selection screen for easy identification. With no computer logging system, the material has to be logged by hand. It is very important that great care is taken in loading or digitizing the material, since these pictures and sound may well become the final material in the project; this is an on-line edit. The sound must be carefully monitored and any problems must be sorted out at this transfer stage before editing. Being a digital sound system, any audio overloads may result in unacceptable glitches or dropout and unusable sound. It is sensible to keep at hand a good test tape, which is known to be correctly recorded, to check against any incoming tapes that might be thought to be faulty. A reference tone at 0 dBvu will correspond to 4 on a European BBC-type peak programme meter which responds to –14 dB on most non-linear editing stations (but this should be checked with the manufacturer's handbook). It is not sensible to alter any of the audio recording parameters when transferring sound into the picture editing system; this particularly applies to frequency equalization.

Do not alter equalization because:

- Any equalization added may not match later when shots are edited together.
- The speaker system in the editing suite is usually not of the best monitoring quality.
- The acoustics in the editing room may be poor and give unreliable results.

Only alter recording level if there is a known problem.

To edit within the non-linear editing system, graphic displays are used; a time line is shown, across which graphical representations of the pictures, audio tracks and video effects pass. The audio display may offer two or more tracks, usually in groups of two for stereo operations. Although up to 24 tracks may be offered, only two or four recording inputs are usual. These can be routed to the various tracks as required within the system. In budget systems it may not be possible to monitor all the audio tracks at once, but this is not a major problem.

Editing is started by selecting and recording the colour bar test signal and the countdown clock with programme information typed in via the keyboard. Material is moved about using a mouse, a keyboard, a trackball or a dedicated controller. The first picture is selected from the clips shown on the selection screen; the clip is viewed and heard and then perhaps trimmed; with the mouse (or other device) it is positioned over the time line which fixes it in place; it can be edited further. The next clip is viewed and the sound is assigned to a track and edited in place next to the first clip. All types of picture manipulation are possible; picture mixes or dissolves are possible selected from a graphic display box. Dissolves can be added providing that there is sufficient overlap of picture material on both the visible A and B video tracks. These effects may take time to produce, for sometimes they cannot be processed in real time, the digital processors taking anything from a few seconds to many minutes to 'render' some special effects. Usually digital sound processing is performed in real time. As the picture edit continues, the soundtracks are built up; well-laid soundtracks can save much time in audio post-production. However, much will depend on the number of tracks available, and whether it is impossible to lay all the tracks required (stereo may need to be recorded as mono to conserve track space) or perhaps only part of a track. Using an EDL will allow the missing sections to be reconformed in the audio editing suite at a later date.

Editing example

The scene to be edited takes place in a restaurant; there are only two tracks available on the workstation. In this example, in mono, additional sounds may well be required to augment synchronous dialogue. These might include sound effects such as a cup being put down, a door being opened and closed or a waitress walking in and dropping a tray of crockery. Should the waitress walk off-screen talking, while other characters carry on with their speeches, the waitress will need to be laid on track 2 and the featured dialogue will carry on, on track 1.

Should a fire bell go off during the restaurant scene and be shown in close-up, the editor will have insufficient tracks. He or she will, therefore, have to leave this effect off, making a note of the time code of the original recording, and the number of the appropriate tape (either manually or via an automatic EDL). The sound of the bell will then be added later, in the audio post-production suite, directly from the original master.

Other sounds may also need to be laid in later because of limited track space. These would include the 'heads and tails' or handles of shots; that is, the early and late parts of shots that were discarded in the picture edit. For example, in the scene a car is passing, and although the car may only be seen for a few seconds to complete the scene, it must be heard for much longer –

that is, as it enters and leaves the scene. Therefore handles (heads and tails) are added to the edited picture shot. Sounds which continue before or after a picture edit are called split edits or L cuts. It is important that any additional sounds that are required from the master camera originals are well logged at the picture editing stage, although the edit decision list will contain much of the information required.

This production has been shot discontinuously with one camera; if it is an episodic drama or soap opera it may well have been shot with two or more cameras in a studio, this being a compromise between speed and quality. Here the video and sound may be recorded straight onto a hard disc recorder ready for immediate editing in a compatible workstation. In a studio there are extensive vision and sound mixing facilities and the sound mixer will produce a more complete soundtrack which requires less sound editing. In the restaurant scene, each sequence would be shot continuously using two cameras. The mixed sound would then include all the voices ready mixed and also the sound of the fire bell at the appropriate time. Some additional sounds would be added in audio post-production, perhaps the clink of cups, featured footsteps and background music.

Editing separate sound

It is possible for a video production to be shot with separate sound, using time code to synchronize together the camera and separate audio recorder. This type of separate sound system tends to be used in situations where a multi-track sound recorder is essential, as in musical programmes; usually a 'rough mix' is produced for editing. This soundtrack will be synchronized with its appropriate picture within the editing system using time code. Alternatively, a sound guide track is recorded on the master video recorder in the field, for use in editing. Later, the master soundtrack is synchronized to the edited master in audio post-production.

Well-laid soundtracks can save much time in the audio post-production suite. In particular, care should be taken as to which of the workstation's audio tracks the edited sound should be sent to. It is important that the sound should move to an alternative track (off-laid) when the incoming signal:

- requires audio equalization changes;
- has a disturbing increase in background level on a cut;
- changes scene;
- has a high level of incoming or outgoing sound against a previous or following low-level sound;
- requires special effects such as reverberation etc.

Once the picture edit is finished, the soundtrack must be mixed. This can be completed inside the non-linear editing system, mixing the tracks within the digital domain ready to be copied out onto the video recorder. The soundtracks can be mixed as they are copied out onto the video recorder; alternatively, the raw tracks can be copied out ready for the audio post-production suite. Various time code options are available at the output; for audio post-production the original camera master time code is a useful addition.

The output from the non-linear editing suite might be:

- A completed video tape with mixed audio track, digital or analogue.
- A video tape with separate unmixed raw tracks ready for post-production, digital or analogue.
- An EDL for later use to conform sound (and picture) in the audio post-production suite (which should, if possible, be provided).
- A data file on disc of all the digital information ready to load into a compatible workstation.

Various standards are available for the interchange of digital data. When the first workstations were developed, manufacturers used many different techniques and formats. This helped to guard their own position in the market, but did not help producers wishing to interchange audio, video and edit lists between workstations. In 1994 the first interchange standard between systems was announced, the Open Media Framework Exchange (OMF). More recent is the AES 31 standard. Manufacturers try to maintain compatibility with their colleagues to help increase the use of computer editing systems generally.

The process of copying the video and sound from the edited master into the digital audio post-production workstation is called off-laying.

When video copies are made for use in the audio suite, it is normal practice to record a copy of the time code in the picture area copied from the edited master. The picture may also have 'burnt into' it (a window dub) the time code of the original camera masters. The time code provides information for cueing and searching, and will accurately display codes when the picture is in the still frame mode. It is used to synchronize the video and the digital audio workstation.

Alternatively, a production can be mixed to the edited master, playing off a high-quality video machine. There will obviously be no burnt-in code in the picture area. Time code can be injected into the picture from the recorder's time code track through a time code inserter, similar to the one used to produce burnt-in time code on copies. This inserter must be capable of reading time code at high speed, and will be slightly inaccurate at very low speeds if it is fed from the longitudinal time code track.

Whatever the type of video player used in the audio post-production suite, the sound has to be off-laid and built up in the digital audio workstation, allowing other sounds to be added before the final mix takes place.

The programme material now consists of a picture of reasonable quality, locked to a digital audio workstation ready to prepare soundtracks for mixing and 'sweetening'.

Off-line editing

Editing video pictures using high quality broadcast equipment throughout the editing process is called on-line editing. Non-linear editing suites are capable of reaching these technical standards.

To reduce costs and increase efficiency, sometimes a different approach is made. The initial edit is produced on a low-cost linear editing suite. Copies of the master tapes are made onto an inexpensive domestic or industrial

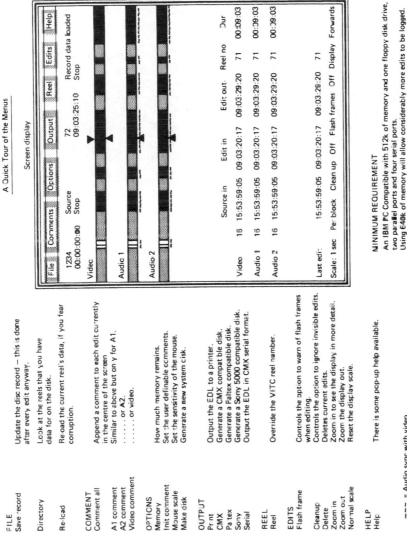

Figure 8.7 An off-line editing menu showing video and two audio tracks, displayed as part of a frame-accurate VHS off-line editing system (JVC) (*Courtesy of JVC Professional Products (UK) Ltd*)

format which includes 'burnt-in' time code from the original source tapes. Editing decisions are made using this low-cost format and an edit decision list is produced to later conform the original masters. This is called off-line editing.

The simplest form of off-line edit is just a single domestic video recorder; using time coded copies of the master source tapes, a manual log is produced which forms the basis of the on-line edit.

At the other end of the spectrum is the use of a non-linear editing suite of low picture quality. From the edit an EDL is generated for later autoconforming.

The more facilities offered within the off-line editing suite, the more expensive the system becomes. The developments in the picture quality of non-linear editing systems means that for video productions, at least, off-line editing will be less and less necessary. If the end product is to be projected film, the situation is different. After a film has been shot and transferred into the editing system, it is necessary to turn the non-linear edit list into a camera negative list. Instead of an EDL that contains time code numbers of the master video, the EDL will contain instructions to the negative cutter showing which frames of the negative to cut.

9

Sound effects

Whatever type of audio post-production system is used – film, video or digital workstation – the basic steps to prepare the soundtrack are the same. First the picture is viewed and reviewed. The sound available from the picture edit will consist of, perhaps, synchronized dialogue, music that has been already cut to picture, and some sound effects that were recorded synchronously on location. It may prove necessary to reconform the sounds already on the sound tracks, not because of quality but because of the limited track space available to the editor. Perhaps the stereo sounds have had to be laid in mono or the heads and tails, the handles, required just couldn't be added. Hopefully much of the information is on the notes or the EDL which will accompany the project. Depending on the system used, the number of tracks available can vary from four to 24 or more.

After the first viewing of the picture, notes are made about the sounds that will be needed for the final soundtrack, and sounds that have already been laid and are available from the source recordings. The analysing of the project material that has been edited is called track listing or spotting. There are two different ways to spot a project:

1 starting at the top and working to the end, noting cues for all sounds required: dialogue, narration, music, effects, etc., or
2 by noting only one particular type of sound or cue as the material is viewed, and repeating the process for each different type of sound, e.g. first viewing – dialogue cues, second viewing – effects cues, etc.

Additional sounds needed will come from the original recordings made on location, from effects libraries, and from music libraries. Some may come from recordings of dialogue, music or effects especially made for the production; some will be in stereo and some in mono. In stereo it will be important to know whether the A B system or the Mid and Side (M & S) recording system was used.

On very large-scale theatrical productions, track laying will begin when picture editing is still going on, and itself will be divided among various groups of editors – working either on particular sounds or on particular reels.

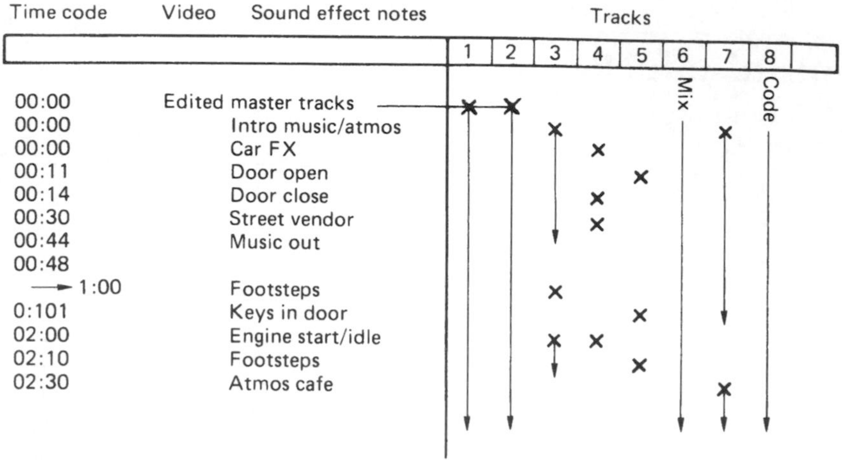

Time code	Video	Sound effect notes	1	2	3	4	5	6 Mix	7	8 Code
00:00	Edited master tracks		x	x						
00:00		Intro music/atmos			x				x	
00:00		Car FX				x				
00:11		Door open					x			
00:14		Door close				x				
00:30		Street vendor				x				
00:44		Music out								
00:48										
1:00		Footsteps			x					
0:101		Keys in door					x			
02:00		Engine start/idle			x	x				
02:10		Footsteps					x			
02:30		Atmos cafe							x	

Figure 9.1 An audio post-production spotting chart

Other sound operations such as automatic dialogue replacement and scoring will also have started.

An audio post-production spotting chart is shown in Figure 9.1.

Once the list of effects, dialogue and additional sounds has been made and collected, they can be laid down in the workstation's directory ready to start track laying. Sound effects will be selected from a variety of sources.

Sound effects

Sound effects occur in any type of television or film production. Props are moved and make noises, doors open and close, actors move about as they perform, winds blow and waves crash on beaches. These effects give the soundtrack atmosphere; they reduce the pauses between dialogue lines and, hence, speed up the action.

Sound effects should be used not only because they obviously appear in the picture, but also because they can become a valuable part of the sound picture. A distant train and hooter can, for example, add space to an open landscape. Every location has its own particular atmosphere. Even in a room with a dead acoustic, the rustle of clothes can be heard, perhaps over the spot effects from a tea party. These can be added in the track-laying process and during the final mix.

Many of the sound effects needed for a production will be provided by recordings from the location, some separately for 'fitting' in audio post-production, and some as part of the picture's soundtrack. Some of the effects will be chosen from a sound effects library.

A sound effects disc covering various subjects from the Sound Ideas Library is shown in Figure 9.2.

Sound effects libraries can be purchased complete in various forms. They have improved immeasurably over the past few years, particularly with the introduction of the portable DAT recorder. Recently, large Hollywood studios

TITLE & DESCRIPTION

A-C

01	Adding Machines, Addressograph
02	Airplanes – Biplane, Jets, Comedy
03	Air, Alarms
04-05	Animals – Bee, Cat, Chimp, Cows, Dogs, Horses, Lion, etc.
06	Answering Machine, Auto-Driving
07-09	Auto – Driving, Skid, Crash, Doors, Race
10	Bells – Fight, Hand, Jingle
11	Birds – Crow, Loon, Nightingale, Robin
12	Boat – Outboard
13	Bottles
14	Bugle
15	Bus, Buzzer, Cable Car
16	Cameras
17	Cans
18	Cash Registers, Chain Saw
19	Clocks – Alarm, Cuckoo, Grandfather, Ticking
20-27	Comedy – Accents, Boings, Crying, Music, Pop, Yells, etc
28	Crashes – Glass, Wood
29-30	Construction
31-34	Crowd – Applause, Boo, Chant, Cheer, Party, Reactions, etc.

D-H

35	Doors
36	Drum – Indian Drum
37	Electricity – Sparks
38	Explosions
39-42	Fire – Fire, Fire Engine, Extinguisher, Fireworks
43	Footsteps
44	Frying, Geiger Counter
45-48	Guns – hand, Shot, Machine, Anti Tank, Howitzer
49	Heartbeat
50	Helicopter
51	Horns – Auto, Truck, Air
52	Horror – Music, Crunch, Stab
53-57	Human – Baby, Burps, Cough, Cry, Eat, Farts, Kiss, etc

TITLE & DESCRIPTION

I-S

58-59	Industry – Hydraulic Machine, Steam, Ambience
60	Jet Ski
61	Jungle Ambience
62	Knife, Lawn Mower, Microphone
63	Morse Code
64	Motorcycles
65	Noisemaker
66	Office Ambience
67	Pager, Spray Paint, Pinball, Printer
68	Projectors
69	Punch, Punch Clock, Rip, Rope
70-71	Sci-Fi – Space Ships, Doors, Pinging, Lasers
72	Slap, Slot Machine
73-76	Sports – Archery, Basketball, Billiards, Curling, Golf, etc.
77	Squish, Stampede, Steam
78	Stereo – CD Player, Record Scratch – Stone
79	Stopwatch, Suction Cup, Sword
80	Synthesizer

T-Z

81	Tap Dancing, Tape Recorder
82-84	Telephone – Rings, Pay Phone, Internal Ring, Tone
85	Ticket Dispenser, Toaster
86	Toboggan, Toilet, Tool Belt, Towel Dispenser
87	Traffic Jam
88	Train
89	Tree Falling
90	Tuning Radio, Typewriter
91	Umbrella, Velcro
92	Video Cassette Recorder
93	Volcano
94-95	Water – Falls, Lake, Ocean, Underwater, Rock Splash
96-97	Weather – Rain, Thunder, Wind – Whip Crack
98	Whistle – Steam
99	Wrench, Zipper

Figure 9.2 A sound effects disc covering various subjects from the Sound Ideas Library (*Courtesy of De Wolfe*)

have offered their own libraries on CD. The Hanna-Barbera library includes effects from such cartoons as the Flintstones, Huckleberry Hound and many more. Warner Bros, 20th Century Fox and Universal offer similar libraries of effects from their productions. In Britain the BBC library of sound effects has been available commercially for 25 years and is one of the largest collections, with over 40 compact discs. Both paper and floppy disc databases of sound effects collections are often available.

A sound effects disc, 'Communications', from the De Wolfe Library is shown in Figure 9.3.

Highly specialized mini libraries are also available, offering such effects as automotive sounds, aircraft sounds and footsteps. These are ideal for manipulating in workstations. Some libraries offer only atmospheres and at realistic lengths (three minutes and more) making them suitable for speedy operations. Libraries can be stored in inexpensive CD jukebox players for instant access to the discs.

Most audio post-production studios will hold and collect their own effects library. The more useful effects are ideally kept within their audio workstation's own library.

Formats for sound effects

Compact discs

Compact discs are digitally recorded, and can provide excellent 16-bit 44.1 kHz sampling quality. They can be started remotely. Access to tracks on the CD is not instantaneous; initially, it is necessary for the reproducing laser arm to move to the track position on the disc. However, once the track point has been found, cueing is almost instantaneous. Professional machines provide shuttle facilities to find cues, although the quality of the sound, in shuttle, may be somewhat limited. Any machine must be capable of accessing as many tracks as there are on the disc; this usually means up to a hundred. The discs can hold up to 60 minutes of material and take up only a small amount of physical space. It is possible to record CD discs (CDRs); the blank discs cost very little and many workstations provide interfaces at small cost. CDRs provide an excellent way of building up a sound effects library, with good compatibility and archival properties.

CD-ROM

Many manufacturers provide sound effects on CD discs. Some are in CD-ROM form, giving the opportunity to provide additional data. Files are provided on compatible formats giving details of looping and cutting points. The database can be printed out.

Digital audio samplers

Digital samplers were originally designed for musicians, to allow musical sounds or notes to be sampled, electronically synthesized, looped and then played through a keyboard with pitch changes. These devices record sound on

COMMUNICATIONS

DWSFX/CD001

No.	Description	Index
1.	GENERAL COMPUTER ROOM ATMOSPHERE	1.05
2.	DATA PROCESSING	0.57
3.	SIMPLE TIME CODE DATA	0.52
4.	COMPUTER DATA BEING TRANSMITTED	1.12
5.	FLOPPY DISK DRIVE – LOADING PROGRAMME	1.00
6.	DOT MATRIX PRINTER	0.25
7.	TELETYPE MACHINE WITH COMPUTER BACKGROUND	0.54
8.	DAISY WHEEL PRINTER	0.31
9.	SWITCHING ON HARD DISK DRIVE – DISK LOADING	1.11
10.	THERMAL PRINTER	0.42
11.	LINE PRINTER	1.04
12.	COMPUTER KEYBOARD	0.44
	COMPUTER KEYBOARD	12.2 0.42
13.	HOLDING KEY DOWN ON COMPUTER WITH BLEEPS	0.14
14.	MORSE CLICKS – 2 DOUBLE AND 1 SINGLE	0.05
15.	COMPUTER ROBOT TALKING GIBBERISH	0.51
16.	ROBOT TALKING 1, 2, 3, 4, ETC.,	0.05
17.	ROBOT TALKING – MON, TUES, WED, ETC.,	0.06
18.	LARGE BUSY OFFICE ATMOSPHERE	10.4
19.	MEDIUM SIZED OFFICE ATMOSPHERE	0.58
20.	SMALL OFFICE ATMOSPHERE	1.04
21.	SWITCH ON ELECTRIC TYPEWRITER	0.05
22.	TYPING FAST ON ELECTRIC TYPEWRITER	0.55
23.	TYPEWRITER – TABBING – ELECTRIC	0.08
24.	ELECTRIC TYPEWRITER – PAPER INTO MACHINE	0.12
25.	ELECTRIC TYPEWRITER – TYPING AT MEDIUM SPEED	0.51
26.	ELECTRIC TYPEWRITER – TYPING FAST	0.58
27.	ELECTRIC TYPEWRITER – SELF CORRECTION KEY	0.23
28.	ELECTRIC TYPEWRITER – CARRIAGE RETURN	0.09
29.	ELECTRIC TYPEWRITER – REVERSE INDEXING	0.10
30.	ELECTRIC TYPEWRITER – AUTOMATIC UNDERLINING	0.17
31.	ELECTRIC TYPEWRITER – PAPER OUT OF MACHINE	0.08
32.	INSERTING PAPER INTO MANUAL TYPEWRITER	0.14
33.	PAPER OUT OF MANUAL TYPEWRITER	0.10
34.	PAPER OUT OF MANUAL TYPEWRITER – FAST	0.02
35.	TYPING AT MEDIUM SPEED ON MANUAL TYPEWRITER	0.58
36.	TYPING POOL	1.16
37.	TYPING A TELEX	0.21
38.	RECEIVING A TELEX	0.17
39.	TEARING OFF A TELEX MESSAGE	0.03
40.	SENDING A FAX	0.53
41.	RECEIVING A FAX	0.30
42.	TELEPRINTER ROOM ATMOSPHERE	1.01
43.	ELECTRIC ADDING MACHINE WITH PAPER ROLL	0.45
44.	FILING CABINET, OPEN AND CLOSE	0.19
45.	PHOTOCOPIER	0.51
46.	SELLOTAPE DISPENSER	0.15
47.	STAPLER	0.12
48.	NEEDLE STUCK ON END OF RECORD	0.49 / 48.2 0.51
49.	RECORD CRACKLE	0.44
50.	VIDEO MACHINE – VARIOUS MODES	1.02
51.	VIDEO MACHINE – VARIOUS MODES	0.43
52.	REWIND VIDEO CASSETTE	0.18
53.	COMPACT DISC PLAYER – DRAWER OPEN & CLOSE	0.07
54.	REEL TO REEL TAPE RECORDER – REWIND	0.34
55.	REEL TO REEL TAPE RECORDER – REWIND & TAPE FLAPS	0.41
56.	CASSETTE MACHINE – VARIOUS MODES	0.22 / 56.2 0.07
57.	CASSETTE, INSERT & EJECT	0.38 / 57.2 0.24
	SHORT WAVE RADIO TRANSMISSION	0.37
	SHORTWAVE RADIO TRANSMISSION WITH MORSE CODE	1.07
58.	RADIO – SHORT WAVE MODULATIONS	0.33
59.	TUNING THROUGH RADIO STATIONS	0.30
60.	TUNING THROUGH RADIO FREQUENCIES – SLOW	0.46
61.	RADIO INTERFERENCE	0.30
62.	OFFSET LITHO PRINTING PRESS	0.16
63.	PAPER FOLDING MACHINE	1.01
64.	CASH DISPENSER	0.56
	CASH DISPENSER	64.2 0.16
65.	INTERNAL WORKINGS OF TELEPHONE EXCHANGE	1.01
66.	TRIM PHONE RINGING AND PICK UP	0.56
67.	TRIM PHONE RINGING	0.17

Figure 9.3 A sound effects disc, 'Communications', from the De Wolfe Library

floppy discs and then convert it to RAM within the machine for playing and manipulation. The looping and pitch-changing facilities of these devices have proved powerful tools in audio post-production. It is, for example, possible to store the sounds of various footsteps in a machine, and by operating a keyboard, make each key-press reproduce a footstep, its pitch being determined by its position on the keyboard. The pressure of the fingers on the keyboard affects the volume of the sound. More sophisticated keyboards and samplers allow various sound effects to be grouped onto one key. The use of these devices can considerably speed up the track laying of spot effects. Not only can footsteps be laid, but also doors opening and closing, balls hitting bats, clothes rustles, wheel skids, and similar effects that require a series of spot effects to be laid exactly in sync. By moving up and down the keyboard, the sound of an effect can be changed significantly; a small car can become a truck, small bells can become large bells, etc. Most devices interface with audio digital workstations using Midi code.

Audio cartridges (MiniDiscs)

MiniDiscs have replaced the old standard NAB (National Association of Broadcasters, USA) audio loop cartridge. They store effects for instant access They can be recorded locally in the studio; access is good since each cartridge can hold its own specific effect. They are ideal for 'dubbing on the fly' where instant access is required; they can also be remotely cued. Each disc can be individually electronically labelled with the legend appearing on the LCD display screen on the machine. It is claimed they can deliver 1 million plays before damage occurs to the disc. They are used for dubbing news or fast-turnaround productions. In this situation it will be necessary to sound mix continuously, only occasionally stopping. Formats need to be easy to use with fast access. Most suitable as sources for sound effects are MiniDisc machines.

Archive formats

Analogue tape

The quarter-inch tape has long been the traditional method of holding sound effects recordings, offering almost instant start. Tapes can be striped with time code, making access easier; it is merely necessary to dial in the time code relating to the sound effect for an automatic location. Although access is not instant, it is easy to lock into other systems if time coded. It is available in various formats and two recording standards, NAB and CCIR (European).

Vinyl discs

Records were once a major source of sound effects material: they are easy to handle and have good access. More than 30 minutes of material can be recorded on one side of a disc and if the tracks are well spaced, it is easy to find a cue. However, the vinyl disc suffers from wear after continuous use and is now used only to play off 'archive' material. Being analogue, quality is not high and depends on usage.

Sound effects and the workstation

The simplest way to manipulate sound effects for audio post-production is to transfer the material into a workstation and move it around within the digital domain. Effects should, if possible, be transferred using the digital inputs of the systems. More useful effects can be stored permanently in a workstation's library rather than just for a specific project.

Material for a project must be well labelled for easy recall. It should be logically divided into groups and subgroups, which can be specially labelled for the project. Sound effects, for example, could be grouped as animals with this divided into birds, farm animals, domestic animals, exotic animals, etc. Automobile programmes would use headings as particular engine sizes, makes with subheadings as exhaust noise, skidding, gear changes, horns, etc. The effects might be 'cleaned up' to remove clicks or any extraneous sounds that could distract from the effect.

Each sound is recorded at full modulation so as to reduce any possibility of background noise being introduced from the system. These sounds are now readily available for any point on the edit, but since they are only copied onto the time line they are not removed from the library or destroyed.

At this point the workstation is full of usable sound and now memory space will only be used up in providing data information as to the actual position of the sounds within the completed edit (and some processing).

A selection of sound effects and music libraries are detailed in Table 9.1

Table 9.1 A selection of sound effects and music libraries

Library	Country of origin	Type
Bainbridge	USA	Effects
BBC	GB	Effects
Carlin	GB	Music
Chappell	GB	Music
De Wolfe (Est. 1940)	GB	Music & effects
Digiffects	Sweden	Effects
Dorsey Productions	USA	Effects
Hanna-Barbera Sound effects	USA	Effects
Network Music	USA	Music & effects
Soper Sound Library	USA	Music & effects
Prosonus	USA	Music & effects
Valentino (Est. 1932)	USA	Music & effects

10

Film editing

Although this chapter is called film editing, almost no editing on film now goes on. Film is used as an acquisition medium for location shooting and as a projection medium for theatrically released films (although even this can be replaced by remote digital distribution of video through cable). Between these two points, film travels a path that all audio-visual programme material passes through.

Film editing is a non-linear editing system; it allows the physical cutting and splicing of film, giving complete freedom to move shots around as required. However, the actual logistics of handling film and in particular storing and logging the material is in itself time consuming and tedious. These problems apply to both picture and sound cutting. The high cost of sound film, which can only be used once, and the advantages of non-linear computer editing have in many cases made film sound editing no longer viable. Figure 10.1 shows editing tables of the types favoured in Europe and the USA.

A film path is likely to be either:

1 film to non-linear picture editing workstation and digital audio workstation to video transmission, or
2 film to non-linear picture editing workstation and digital audio workstation to film release.

Audio workstations are now preferred for track laying rather than sprocketed film. Appendix B gives an overview of traditional film editing.
Using traditional methods:

● There was no check of technical quality while laying the tracks.
● It was impossible to change the quality or level of the sound at the track-laying stage.
● It was difficult to judge the suitability of sounds whether stereo, mono or with noise reduction because of the limited quality of the cutting equipment.
● Since there were no recording facilities in the cutting-rooms, it was time consuming to change audio.

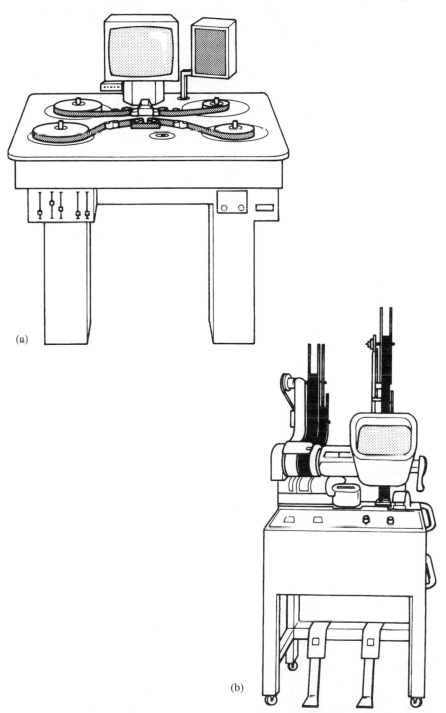

(a)

(b)

Figure 10.1 (a) A flat bed editing table as favoured in Europe and (b) an upright editing machine as favoured in the USA

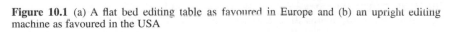

- Although soundtracks should not lose sync, when this did happen, there was no specific time code reference to match picture and the soundtrack (unless the film and soundtrack had already been rubber or frame numbered).
- When laying sound, only two or three of the tracks laid could be heard at one time.
- Logging sound was very time consuming.
- Eventual sound quality was good but sprocketed magnetic film can suffer from physical damage from use, causing high frequency drop-off.
- It was labour intensive.
- The material is bulky to store.
- Magnetic stock is expensive (and becoming more so) and cannot be used again.

In the early 1990s, comparative tests were made between traditional film sound editing using sprocketed magnetic film techniques and those using an audio workstation. A 1000 ft (12 minute) reel of 35 mm picture was chosen from a major studio production about to be laid with sound effects, and an analytical comparison was made of the two methods. Speed, efficiency and cost were analysed:

- Spotting, preparation: Eighty-four sound effects were created, taking four person-days using traditional methods. The digital method took one day with material savings of 35 per cent. Using the digital system there was a marked reduction in transfer problems, less cutting-room downtime and increased quality.
- Editing, track laying: Traditionally 13.5 person-days produced the 61 tracks required. Digitally, three person-days produced the 31 tracks the workstation needed, thus producing a material saving of 15 per cent. With the digital system there was the ability to play back and complete scenarios, and combine effects onto fewer tracks – with equal efficiency and with the ability to call up effects at any time.
- Mixing: Traditional track laying had created 61 tracks, the digital system 31. Loading traditional sprocketed machines took 10 to 15 minutes, but with no time required for the digital mix. The digital schedule was faster by 12 per cent, with the advantage of a visual preview of all the tracks and little noise in the system.

This report showed the economic advantages of using digital systems in both time taken and saving in materials. The ease of track laying meant that with careful planning, fewer tracks were needed for the same job.

Telecine

To edit sound tracks in an audio workstation, the picture needs to be available on video. If the original is on film, this must be transferred to video. Once there, it may never need to be returned to celluloid and no relationship between the camera negative and the video need exist – this is the path of films made for television. Alternatively, if the picture is to be shown in the cinema it is essential to be able to relate each frame of camera negative to each edited frame in the non-linear editing system so that a print can be produced. This process begins at the preliminary stage of transferring the negative to video for editing. A telecine machine is shown in Figure 10.2.

Key numbers/edge codes

Although camera film is not time coded, each frame is uniquely numbered by the manufacturer. This provides vital synchronizing information to match an EDL to a negative. When a photographic negative is manufactured, numbers are exposed along the edge of the film (every 1 ft in 35 mm, every 6 inches in 16 mm) which become visible when the film is developed. These numbers are known as key or edge numbers. Each number uniquely identifies a frame of camera original.

Edge numbers can be read by eye or by a machine (using a bar code reader); they identify frames in relation to the position of the code (e.g. code XXXXX plus eight frames). These machine-readable key codes are invaluable in film projects that will eventually be printed up for theatrical release.

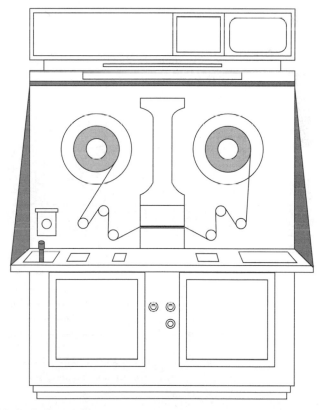

Figure 10.2 A telecine machine

Figure 10.3 shows an example of Edgeprint format for Eastman colour negative films.

The telecine machine converts the camera negative image to a positive video image; it can also grade or alter the colour of the image (in a video transmitted project) at the transfer, although this is more often carried out later in the video domain.

Film to film

In film, to non-linear computer editing, to film projects the editing system's edit decision list (EDL) is used to conform the camera negative to the picture. It is essential that at the start, before any editing begins, each frame of film is identified and related to its frame of video. The simplest way of doing this is to give each roll of film a time code sequence directly related to the edge numbers. Since the film is physically sprocketed, a time start point can be identified at a specific frame; this also becomes the time code start on transfer to video. At the start point (for example 01.00.00.00 time – hours can identify film reels) the key number is read off; as the sprockets pass it is possible to read time and mathematically relate it to edge numbers – all that is necessary is to know the start point time. This is usually identified by punching a hole in the film. To conform the film, the time code readings from an EDL are converted to edge numbers. The negative can then be cut frame to frame against time codes.

Alternatively, as the video transfer takes place, the edge numbers are machine read automatically by the telecine machine and logged and corre-

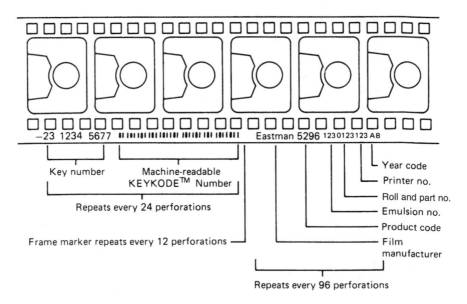

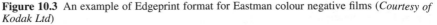

Figure 10.3 An example of Edgeprint format for Eastman colour negative films (*Courtesy of Kodak Ltd*)

lated against the video's time code. This uniquely labels every frame of film. The sync sound for the project is conformed later, using the video transfer as the master. Throughout the picture editing, the original time codes are maintained; on completion of the editing these time codes are tied back to the key codes, allowing the negative to be matched to the video edit frame for frame. In NTSC systems there are additional problems because the pulldown mode needs to be known.

Pulldown

Within the PAL system, film is run at 25 frames per second (fps) – half the mains frequency to which video is tied, a simple relationship. In the NTSC system, the relationship is more complicated. Here the 24 fps film speed has to be tied to a television video rate of 29.97 fps. Mathematically a correlation can be produced to enable copying between the two. Film frames are 'pulled down' and copied first onto two fields of the video tape and then three fields of the tape; these are defined as pulldown modes and may be a 2–3 or 3–2 sequence depending on which fields are duplicated. The picture edit must be tied to one of these sequences if the edited videotape is to identify actual frames on the camera negative. These do not affect the sound synchronization in any way; providing the film negative is now cut correctly against the workstation picture or film cut list, sound will conform exactly to the picture edit.

Film time code

There is a long tradition of recording time code onto audio and videotape, but only some acceptance of recording time code onto photographic camera negative. However, time code does have uses where complicated post-production is involved, and various systems have been developed. It provides specific frame identification for conforming the edited video pictures to the camera negative. A film time code system is shown in Figure 10.4.

Arriflex code system

In this system the SMPTE/EBU time code appears in digital form along the edge of the film. The system uses one LED (light-emitting diode) either in the gate or within the magazines – where it has to compensate for the film loop. In operation the camera's time code generator is synched to an external generator also feeding the audio recorder. Arri time code and audio time code are the same and the transfer to video of both sound and camera negative is automatic. At the transfer the R-Dat, Nagra Digital or quarter-inch time coded analogue machine is chase locked to the telecine machine using the film time code. Both picture and sound should have 10-second run-ups. Ideally the sound recordings should follow the order of the camera shots and there should be only synchronous material on the sound rolls, with a second recorder used for wild tracks and guide tracks. Each picture roll should have its own sound roll. This means that the sound will follow the camera negative

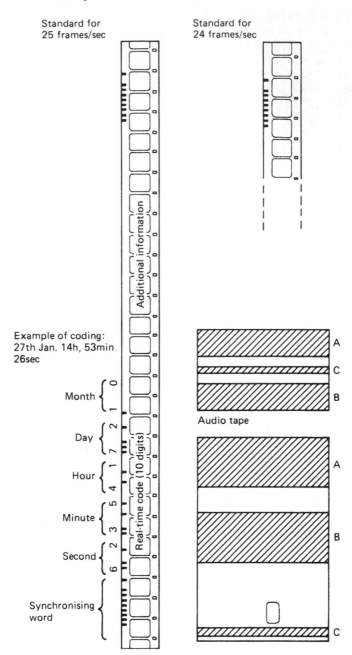

Standard for
25 frames/sec

Standard for
24 frames/sec

Additional information

Example of coding:
27th Jan. 14h, 53min,
26sec

Month

Day

Hour

Minute

Second

Synchronising
word

Real-time code (10 digits)

Audio tape

A

C

B

Magnetic film
Track assignments are:
(A) and (B) programme material
(C) time code

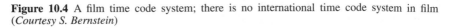

Figure 10.4 A film time code system; there is no international time code system in film
(*Courtesy S. Bernstein*)

at the first attempt of autoconforming; otherwise many rollbacks of camera negative by the telecine machine to try to follow the sound may scratch the valuable negative. As this takes place, the Arriflex time code derived from the negative and the manufacturer's own code identifying each frame of picture is held in a computer database together with the videotape time code.

When the video edit is completed, the time code on the video is referred back to the film manufacturer's code, producing a neg or cut list. The negative is then matched up and printed to produce the final print with its time code providing confirmation of print to video synchronization.

Aaton system

The Aaton system was specifically designed for 16 mm use. Both digital and visual-readable numbers are printed onto the film stock in the camera using a row of seven red light-emitting diodes (LEDs). The system has been developed further for use with 35 mm film. Information recorded includes time code in digital form with visual information covering frame, scene, take, roll, production, equipment number and the date of the shooting.

In an extremely clean environment the negative is now prepared – the shots are identified from the list and then cut together with any special visual effects required. The completed negative is then ready for printing with the soundtrack.

Photographic film sound recording

In the cinema, the sound is usually reproduced from an optical soundtrack, photographically recorded on film. Optical recording is a highly specialized form of sound recording which is met in audio post-production for the cinema.

First, the picture negative is prepared and an intermediary copy is now made, an interneg, of both picture and sound. The cut camera negative is run through the printer and then the soundtrack negative is run through against the same piece of stock, producing on processing a combined print.

To record sound photographically, light-sensitive film is drawn across a slit which is formed by two metal ribbons held in a strong magnetic field. This 'light valve' opens and closes as signals are fed into it. The light shone through this slit from an intense quartz halogen lamp traces a path onto the photographic film which varies in intensity with the signals. An optical recording system is shown in Figure 10.5.

This standard optical track is located near the edge of the film, just inside the sprocket holes (35 mm). When the film is processed and printed the photographic image of the soundtrack is exposed as a white-on-black line of varying thickness. Since the areas of black and white within the soundtrack are continually varying, this is called a variable area soundtrack. It is now almost universally used in optical sound recording.

The soundtrack is replayed by being projected, not onto a screen but onto a photoelectric cell, which picks up the variations in light and converts them into electrical signals. This reader must be well set up otherwise the bandwidth of the system will suffer.

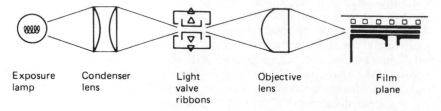

| Exposure lamp | Condenser lens | Light valve ribbons | Objective lens | Film plane |

Figure 10.5 An optical recording system used to produce a variable area soundtrack. The light valve ribbons move up and down within the magnetic field

The photographic soundtrack does not reach the high fidelity standards of a digital soundtrack. However, analogue recorded optical tracks are of reasonable quality, and much of the poor sound associated with them is due to the way in which they are handled, rather than any major defect in the recording system. Many of the audio problems relate to high noise levels, which increase with use; a film soundtrack can easily be damaged as it passes through a projector. When the clear areas of the film have become scratched, they are picked up by the reproducing photo-electric cell as clicks and background hissing. Noise reductions techniques can reduce these problems dramatically.

The optical soundtrack has a frequency response that is substantially flat up to 12 kHz (only an adequate figure), equal to that of 14-bit digital sound with a 32 kHz sampling frequency. However, it is substantially free from compression effects at high level, and at modulations of up to 100 per cent performs in a perfectly linear manner; beyond this, like digital audio recording, it can go into heavy distortion.

The noise in an optical soundtrack is mainly in the mid-range region, caused by the granular structure of the film, and since the early 1930s noise reduction has been applied. It is described as 'ground noise reduction' and its aim is to reduce the transparent area of the track to near zero, when there is no signal. Normally, a solid state delay device is used which retards the audio fractionally, allowing the noise reduction system to 'open up' just ahead of the required programme audio (anticipating noise reduction).

Handling noise, grain noise and replay sound noise are all proportionally reduced as the width of the track is decreased. When the track is 100 per cent opaque, no noise is sent through the system, although in practice the track never completely disappears and a 'bias line' remains. Modern optical tracks are recorded with a bias line of 2 milliseconds. With suitable ground noise reduction, noise levels as low as 70 dB below 100 per cent modulation can be achieved. The aim is to achieve maximum recording level.

To assist the recording engineer in achieving a maximum level without distortion on the optical soundtracks, two indicator lights are often provided on a recorder in addition to meters; one flashes yellow on a clash of the light valve, indicating 10 milliseconds of distortion – which is considered inaudible; one flashes red on clashes over 100 milliseconds, indicating audible distortion.

Photographic recording test procedures

To ensure that the film sound is exposed successfully, 'sensitometry tests' are made through the recording chain so as to determine:

1 the optimum exposure or the amount of light used in the optical camera;
2 the processing time needed to produce the right contrast, creating good definition between the dark and clear parts of the film, both for negative and print material.

Owing to the nature of the photographic process, it is impossible to obtain the ideal 100 per cent white and 100 per cent black intensity on a film soundtrack, so a compromise has to be reached which will produce good sound quality and low background noise. When the sound is printed onto the release print, the image should have exactly the same definition as the original camera recording. However, the system is not perfect and there is likely to be some image spread, which will result in distortion and sibilance. To optimize the processing and printing conditions, film 'cross-modulation tests' are carried out.

A combined audio signal of 400 Hz and 9 kHz is recorded at various negative exposures.

Prints are produced and played off on a cross-modulation analyser. The best print will produce the highest output from the high frequencies.

Stereo optical soundtracks are now found on most motion picture film releases. The alternative is a separate synchronized audio player or magnetic stripe. A magnetic stripe is applied after the picture has been printed. It has been estimated that it costs ten times as much to produce a magnetic print as it does an optical release print so they are now no longer used, having been replaced by digital sound. This can either be directly recorded onto the film as an optical soundtrack or played off as a separate CD-ROM synchronized by optically recorded time code on the film.

The optical recording system is well suited to stereo, since two separate tracks can be easily recorded onto a photographic emulsion. A dual photoelectric cell is used to pick up the two separate outputs. These tracks are often Dolby A or SR encoded, the format being called Dolby Stereo bi-lateral Variable Area, or just Dolby SVA. The use of a sound matrix within the Dolby recording system allows four separate sources to be encoded onto the track. Three are for use behind the screen, left, centre and right, and the fourth, surround channel is located in the audience area. Figure 10.6 shows the Dolby four-channel surround cinema system.

Dolby's digital optical sound system allows six separate discrete digital tracks to be recorded onto the film, screen left, centre and right with sub-bass behind and left and right sound in the auditorium. The system utilizes black and white clear pixels optically recorded between the perforations on the 35 mm film and Dolby's AC3 coding system.

The same speaker layout is adopted by Digital Theatre Systems (DTS) whose system first appeared in 1994. The system uses CD-ROM technology, where the technical quality from disc is better than from film, where the digital information has to be stored in a limited space.

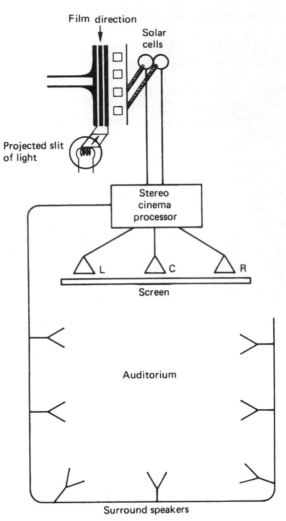

Figure 10.6 The Dolby four-channel surround cinema system

The optical track carries time code data that also contains reel numbers to allow a CD-ROM to follow the film. This code track is photographically printed between the picture and optical tracks. It offers added flexibility since foreign versions may not require new projection prints, merely replacement discs.

In the Sony digital system, blocks of clear and cyan pixels are printed continuously down both of the outside edges of the film. It uses the ATRAC technology found in Sony's MiniDisc recorders. The audio information from the track is decoded to give behind the screen, left, left inner; centre, right inner and right and sub-bass, with again left and right surround in the auditorium, although many installations are simplified to five channels.

Figure 10.7 shows the surround sound systems on film.

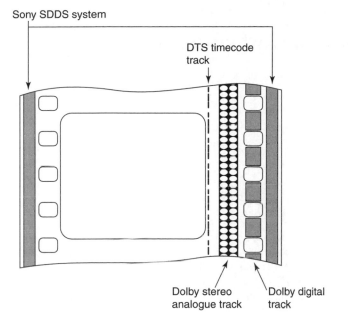

Figure 10.7 The surround sound systems on film

Optical sound is:

- easily copied by the photographic process;
- inherently distortion-free (but goes into very heavy distortion on over-modulated peaks);
- susceptible to poor handling;
- of adequate quality unless digital;
- produced to universal standards;
- entirely mono-compatible in analogue stereo format;
- capable of being produced in multiple formats on any film reel.

11

Audio workstations

The audio post-production industry has widely accepted the audio digital workstation with its speed and flexibility, which follows the versatility of film operations where tracks can be randomly accessed and moved around at will, albeit by actually physically unpicking a splice and remaking it!

The advantages of track laying using a sound-studio workstation are:

- Specialist staff are more expert and quicker at operating sound equipment.
- Specialist equipment speeds up operations and provides facilities which picture-only suites cannot supply.
- Mixing can be fully automated in a specialist sound production suite, saving time.
- With experience, it is possible to start the mixing process while track laying.

The advantages of track laying and mixing in the non-linear picture editing suite are:

- The production stays in one environment with possible convenience.
- The tracks are handled by one person who knows the material. The material does not have to be moved around from editing room to editing room, saving costs and time.

Audio digital workstations (Figure 11.1) record sound digitally on computer discs. They unify all the stages of sound post-production, from holding a sound library to final mixing. They are capable of synchronizing and controlling equipment: storing, recording, editing, mixing and processing sound as well as reproducing it for immediate playback. The facilities offered vary enormously from manufacturer to manufacturer. They take advantage of many different audio post-production techniques but are purely an audio post-production device, not a picture editing tool. They will interface with any film or video format, but are usually locked to the U-matic, Beta SP format, to a disc-type video recorder or to video running within the host computer. Since the speed of the system is dependent on the lock-up speed of the slowest machine (usually the videotape player), instant access video provides the quickest method of operating.

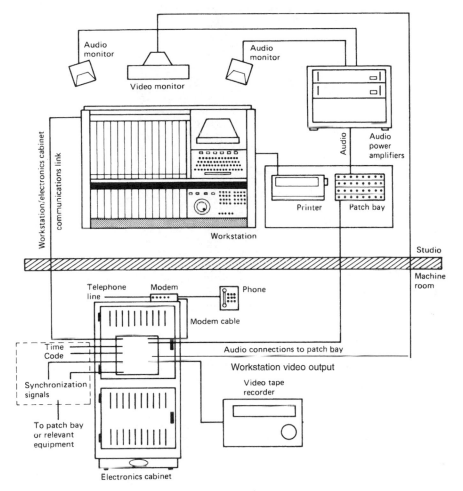

Figure 11.1 A workstation with its control electronics

Audio digital workstations are by their nature non-linear editing devices, and give instant access to any located point in the picture edit or audio held in the library. Although work can be carried out in any order, soundtracks follow a linear picture and are tied to it. The advantage of the audio workstations is the ease with which material can be found and manipulated on the created tracks.

Technically there are two types of audio workstation, dedicated standalone systems and those using standard desktop computers with additional hardware and software. Many desktop computers lack the processing power to handle all the requirements for digital audio manipulation, and manufacturers provide their own software and hardware to produce very versatile units. The host computer acts mainly as a user interface with an audio processing card handling the editing and audio processing.

Dedicated workstations tend to be more expensive and specialized, having been introduced before desktop computers had sufficient power for audio

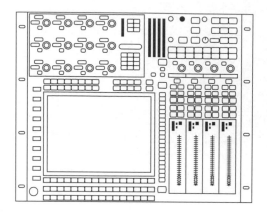

Figure 11.2 An integrated workstation and mixer in a dedicated system (*Courtesy of AMS Neve Ltd*)

manipulation. They tend to be produced by manufacturers of long standing because of this.

Figure 11.2 shows an integrated workstation and mixer in a dedicated system.

Audio workstations offer varying facilities, depending on their concept and their price. Manufacturers have designed their systems independently and the same functions may be called by different names; 'a piece of audio' is to some a segment but to another sound editor it might be a clip, a cue, an event, a take or an item – it depends which system you use of the 50 or so on the market! Some use music-based terms, some video, some film; the following descriptions relate to most, but not all, on the highly active market.

All audio digital workstations offer the same basic facilities; various screens show work in progress.

Cue sheet

Information for controlling the workstation is shown on the video display screen with control from such devices as a keyboard, a touch-sensitive screen, or a cursor via a mouse. The important visual displays are available, individually or together. The graphic display cue sheet is the heart of the system, which might also be called the arrange, assemble or project page. Some cue sheets display a waveform representation of the soundtrack similar to the modulations of an optical film soundtrack (Figure 11.3); some require sounds to be named instead. It is here that the editor lays the sounds. He or she usually selects these from the library page, which holds all the sounds in a directory.

Meter screen

The meter screen helps check sound processing levels such as volume, reverberation or multi-channel panning. As the sounds are played the screen dis-

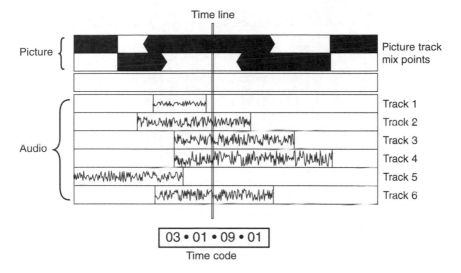

Figure 11.3 Soundtracks as often represented on a workstation screen

plays, in real time, the automation level set for each channel, the curves for frequency equalization and dynamics, etc. The screen may be part of the cue page.

Figure 11.4 shows a workstation meter screen with level controls and equalization.

Library screen

The library screen displays the sounds loaded into the system. They are labelled, numbered, and located by time code information, and are transferred

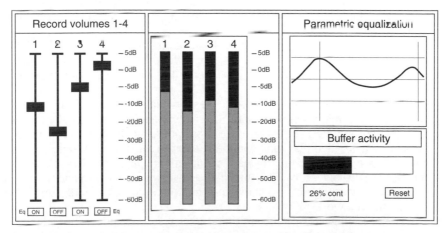

Figure 11.4 A workstation meter screen with level controls and equalization

Job: WIZARD 1		Reel: 1		12:10:20	Hard disk files

Job/reel/trk	Segments	Duration	Date/time	Comments
WIZARD 1			1/05/97 11:25	
Reel 1		00:10:45:00	2/12/97 8:25	
Track 1		00:10:45:00	2/12/97 8:25	
	Bang	00:00:01:20	2/12/97 8:35	
	Crash	00:00:03:25	2/12/97 8:37	
	Boom	00:00:04:27	2/12/97 8:38	
	Footstep	00:00:02:05	2/12/97 8:40	
	Door	00:00:01:17	2/12/97 8:41	
	Beepbeep	00:00:01:02	2/12/97 8:43	
	Smash	00:00:02:07	2/12/97 8:45	
	Jets 1	00:00:03:06	2/12/97 8:48	

Total disk time: 120 track minutes. Time used: 3 reels, 50 minutes used.
Time available: 70 track minutes

Full text:

Input:

REC/PLY	NEWJOB/REEL	CONFIGURE	SEL DRIVE	TRANSFER	LOG IN

Figure 11.5 A library screen

to the cue sheet to be laid as soundtracks. Sounds are are identified by time code or label and triggered to be heard. Once the sound has been found, the library screen is returned to its cue sheet page. The library can be set up with any name of subgroup or group required; it is entirely up to the user.

Workstation facilities include:

- Auto-fill: Automatically calls up the heads or tails (handles) of a segment. It is useful to create crossfades without returning to a directory to find the original audio.
- Automatic deglitching: This is a function on all workstations. When a segment is moved next to another and joined as an edit, there is likely to be a nasty click. This is removed automatically with a very fast automatic crossfade.
- Channels: The number of independent sounds that can be played simultaneously internally; this may be irrespective of the number of actual outputs. Therefore eight channels may only be available at four separate outputs from the machine.
- Crossfades or dissolves: In order to sustain real-time crossfades, some workstations actually support more internal channels than published; two channels are required for a real-time crossfade, and these extra channels are sometimes called playback channels. In non-real-time crossfades, extra channels are not used but instead a one-off calculation takes place to produce the dissolve; however, this takes extra time to perform and uses up more disc space. Figure 11.6 shows a selection of curves of fades out and in offered on a workstation.
- Disc space saving: It is often said that digital systems don't record silence; in fact, they only save silence by removing it from between recorded seg-

ments. Recorded silence between spoken words, for example, exists on the file, just as the actual speech does. However, when laying tracks much of the silent spaces between segments can be removed, and systems will mark out segments and automatically remove the silences to allow an increase in available recording time.

- EDLs: The systems used in non-linear picture editing and audio workstations need to be specified and compatible. They can operate by recording all the material into the workstation and then using the list, or by controlling the outside source machine to follow the list.
- Event: The time a segment is scheduled to play.
- Inputs/outputs: These will be stated as digital and analogue. Analogue inputs are individually mono, but digital inputs will be specified as a particular digital interface, such as AES/EBU which is stereo, or TDIF which has eight dedicated inputs. In workstations the inputs will be much fewer than the outputs; for example, two inputs and 16 outputs on a 16-track workstation. In simple systems, one stereo input is sufficient to allow the workstation to be loaded. But if the system is to be used with a four-track video recorder, this will be insufficient for transferring all the four tracks at a single pass. Problems can arise too if multi-channel music mixes are to be copied into the system; sometimes these are better left in their original formats, perhaps as a modular multitrack dubber synchronized to follow the workstation during a final mix.
- Maximum recording time: If a system cannot record across discs there will be always be a maximum recording time without interruption; this will in effect be the maximum programme length possible to transfer into the system. The figure is unlikely to affect operations.
- Playback assignment: Track assignment will refer to routing and the ability to select any input to any output internally which can speed up operations.
- Slide/reveal/overwrite: Allows a segment to be placed over another to reveal the lower segment as required. The segment is not butted automatically to the nearest segment as is the standard action.
- Split segment: Allows a segment to be split in two for splitting onto another track; this should be a one-button operation. This is ideal for splitting up dialogue onto adjacent tracks to adjust varying background levels; ideally it is possible to glue the segments together again easily.
- Stacking: The number of takes a system can stack is important. Stacking allows takes to be 'recorded over' successive takes without destroying them. These are usually automatically named after the first generic name, i.e. ADR Artiste 01, ADR Artiste 02, etc. These takes can then be listened to and the best retained.
- Storage: A bank of drives can be one or more, but increasing the number of external drives will not increase the number of channels, unless these are multiple drives working in parallel, although this will, of course, increase the storage space. As a guide, 16-bit recording at 48 kHz sampling will use around 1 MB every 11 minutes or GB for every 3.1 hours. If the sound is compressed, this will increase by approximately the compression rate. Figure 11.7 shows the minutes of storage time available on recording discs.
- Synch to offset: Systems are said to synchronize to an offset if they will synchronize to a specified point in a segment rather than the in or out point.

This is used, for example, when synchronizing a 'wild' car passing shot to picture where the middle of the segment has to be synched up.

- Time code: This may be Midi or SMPTE/EBU based. Music-based workstations are often Midi controlled. Time code is locked within the machine to the digital sampling rate; this is a more complicated engineering problem with NTSC systems than with EBU systems. Synchronizing systems that only trigger lock are unlikely to notice discontinuous time code. The parameters for freewheeling through discontinuous time are also significant.
- Tracks: Not always the same as channels. An eight-track system may need 16 internal channels to crossfade through eight tracks in real time. It may also be possible to assign additional sounds as (virtual/background) tracks to a channel, but only eight specified tracks can be played out at once. In some systems one can only view a graphic representation of a limited number of tracks at a time, but record more; for example, view any four tracks at once in an eight-track system.
- Waveform view: Allows the waveform of an audio segment to be viewed; this may speed up finding a particular frame. Figure 11.8 shows a waveform display of tracks on a workstation cue screen

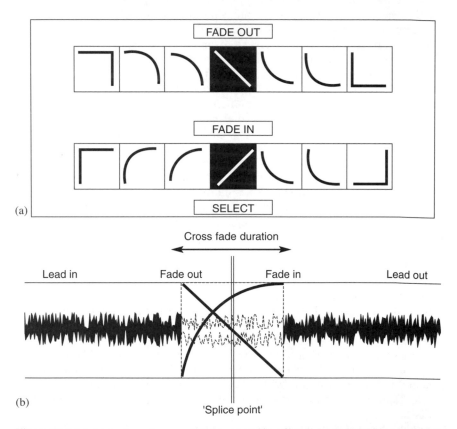

Figure 11.6 (a) A selection of curves of fades out and in, offered on a workstation (simplified); (b) Construction of a cross fade

Disks	48 kHz 800 MB	44.1 kHz 800 MB	48 kHz 1200 MB	44.1 kHz 1200 MB
1	124	135	190	207
2	257	280	390	425
3	390	425	590	642
4	523	569	790	860

Figure 11.7 Track minutes of storage time available on recording discs (up to four discs being held by the workstation at one time) with varying sampling rates

Technical screens

Other screens are often available, giving technical information such as the signal routing and configuration, the synchronization interfaces with time code, the digital sampling rate, buffer activity, diagnostic programs for maintenance and so on.

Defining material for workstation use

The material used to make a project has four distinct names:

1 Project/composition data: This is all the data required to produce a project except the actual audio used.
2 Source audio: This refers to all the audio used in its entirety to make the project.
3 Consolidated audio: The actual audio used to make up the project, usually including the handles.
4 Final master audio: The audio only used to make the project without data but including the time code.

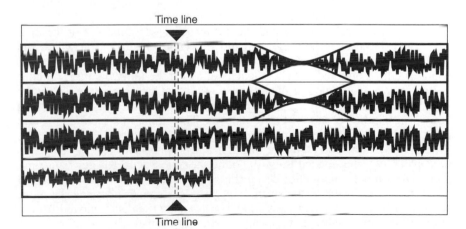

Figure 11.8 A waveform display of tracks on a workstation cue screen

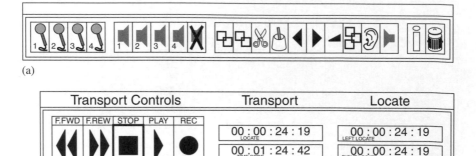

(a)

(b)

Figure 11.9 (a) Part of the main screen containing the toolbar used for most editing functions, specific to each manufacturer. (b) Workstation screen transport controls operated by a mouse; alternatively, the left, right, down and up keys on the keyboard can be used

Using the workstation

Cue sheets

As the track building takes place, the time line display will show a representation of the sounds. In the first instance the cue sheet will display the sound from the edited video master (either as a block or the representation of the waveform) which has been off-laid into the workstation.

Figure 11.9 shows part of the main screen containing the toolbar used for most editing functions, specific to each manufacturer, and the screen transport controls.

The motion controls are of the standard type, fast forward, fast rewind and play and record. With a dedicated controller (Figure 11.10) or mouse it will be possible to jog the audio or video to a specific point with the two in lock to find a cue. Alternatively, a cue can be found by examining a representation of the waveform of the audio and identifying a point.

When 'play' is selected, the cue sheet moves across a time line or the time line which represents the play head moves. This is the point at which the audio is triggered and heard on the monitors. The time code is displayed when sound is edited onto the cue sheet; each sound can be individually labelled.

Like any post-production operation, a degree of preparation is necessary before starting audio mixing. Whatever system is used, the master video or a copy of it will be needed for the sound mix. The audio post-production suite must be capable of handling that format and synchronizing it to the SMPTE/EBU time code track or through a Midi interface. The system must be capable of being loaded up with sound, complete with any relevant time code. The simplest method of doing this is by using a videotape with edited soundtracks ready to be transferred into the system. Sometimes, in more sophisticated set-ups, the audio will arrive as a data file disc removed from a compatible picture editing workstation. Here no transfer is necessary – the disc will play and synchronize immediately with time code locking picture

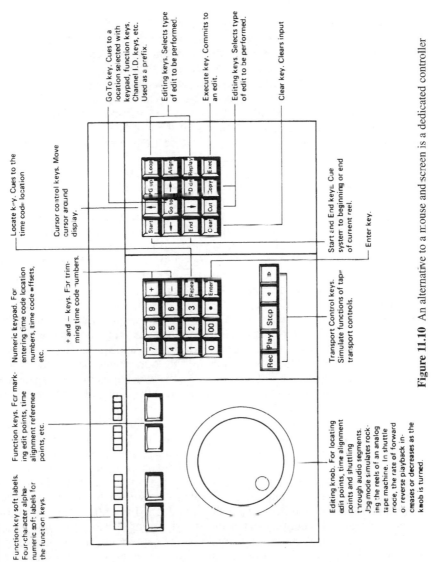

Figure 11.10 An alternative to a mouse and screen is a dedicated controller

Control section

Function key soft labels. Four-character alpha-numeric soft labels for the function keys.

Function keys. For marking edit points, time alignment reference points, etc.

Numeric keypad. For entering time code location numbers, time code offsets, etc.

+ and − keys. For trimming time code numbers.

Locate key. Cues to the time code location

Cursor control keys. Move cursor around display.

Go To key. Cues to a location selected with keypad, function keys, Channel I.D. keys, etc. Used as a prefix.

Editing keys. Selects type of edit to be performed.

Execute key. Commits to an edit.

Editing keys. Selects type of edit to be performed.

Clear key. Clears input

Start and End keys. Cue system to beginning or end of current reel.

Enter key.

Editing knob. For locating edit points, time alignment points and shuttling through audio segments. Jog mode simulates rocking the reels of an analog tape machine. In shuttle mode, the rate of forward or reverse playback increases or decreases as the knob is turned.

Transport Control keys. Simulate functions of tape transport controls.

and digital audio data. Some audio and project data may arrive on other compatible systems such as a data tape which needs transferring into the workstation, but perhaps with the advantage of a high-speed transfer. Sometimes an edit decision list will arrive with original audio or videotapes and edited picture. All these systems can maintain the audio in the digital domain, which is the ideal way of working. At some point in audio post-production it may be necessary to return the signal to analogue, perhaps when using an analogue video recorder or processing the audio through an analogue effects device; here it is important that levels are correct, otherwise noise may build up. Once the audio and the video are loaded into the workstation, additional track laying can begin.

In practice, an edited video programme with 'raw' tracks is the most common form of audio to be presented at the audio post-production stage. If these tracks are played directly into the workstation they can be used immediately. If, however, a low-quality videotape format is provided to view picture and sound, audio will be provided on a separate format such as a synchronous modular eight-track dubber or R-Dat. The R-Dat machine should be a professional synchronous one; a 10-second run-up time should be allowed. The tracks will be transferred or off-laid into the workstation, the workstation following the time code. (The video might be transferred into the workstation's video card as well for viewing.)

If the production is complicated, it may be necessary to reconform all the original sound into the workstation, particularly if some of the audio could not be laid during the original edit though lack of tracks. For this operation, the location source tapes and other time coded material used, perhaps master music tapes, will be needed, hopefully with logged time code. These will be either master videotapes or, for a film production, separate sound tapes. An EDL derived from the picture editing workstation can provide all the necessary information. It will tie up the code on the source tapes (lost on copying in the editing suite) with the time codes used on the master edit. Sophisticated workstations will give the actual offset time code figure needed to synchronize the original videotapes or R-Dat with the edit and then autoconform the two. Complete autoconforming of all the audio tracks recorded in the EDL is possible. The EDL will tie in the time codes to the appropriate tape number. Two methods are possible: by recording all the source material into the workstation and then using the EDL to automatically select and correctly sequence the material, or by controlling the source machine and only recording the selection specified by the EDL.

EDLs can usually only recall four tracks of laid audio. However, the picture editor may make notes to appear on the EDL regarding wild tracks, footsteps or dialogue for fitting as he or she comes upon them in the edit.

Conforming tapes where there is no EDL is more difficult, unless some effort has been made at the picture edit to log the original camera shots used in the master edit. For example, in a simple linear videotape editing suite the recording tape will have been arbitrarily pre-striped with time code to allow the master edit to be built up using the insert editing mode. This time code, ideal for synchronization, will not relate to the camera original time code in any way. Additional information is, therefore, necessary to tie up these two seemingly unrelated codes, one edited master, one original location time code. Trying to find a specific take from a pile of sound rolls with little or no information is a time-consuming and thankless task. Once a specific take is found it should be loaded into the workstation in synchronization speed and carefully labelled under a specific directory. It can then be fitted by matching up the displayed waveform modulations or by listening by ear as the two sounds phase together when played together.

Laying tracks

Initially the workstation cue sheet is selected. This consists of a picture track, with facilities for notes to be written in, together with its audio tracks, which can be labelled appropriately, and the time code start point of the mix.

The sounds are usually grouped for ease of mixing as they are laid – one track containing atmosphere tracks, another track containing spot effects and another, for example, dialogue. Once the tracks have been laid they can be moved around at will, to ease final mixing.

Example of audio editing

The scene to be track laid is of a busy street with cars going by and a nearby street vendor. It was shot silent (MOS/Mute). Like all videotape programmes, the material starts with a clock counting down to the first frame of picture. The programme has already been spotted or effects listed and these exist in the workstation as a separate library for this particular production. Some reconforming of the original location tapes is necessary. First, the original location tapes (video or audio tapes) need to be available in a format suitable for the audio post-production suite. This may mean having the tapes transferred elsewhere or hiring a machine specifically for the particular job. These tapes then need to be tied into the final picture. This is not as simple as it may at first seem, since the master original tapes and the edited master videotape no longer have the same time code. Often only an occasional take needs to be reconformed rather than a complete programme. Here, unfortunately, the edited video master does not include all the sounds that were available from the location tapes; with insufficient track space two original takes need to be conformed from the editor's notes. Three 'handles' are missing on the edited soundtrack. The sound of a car has only been laid for the length of the picture (when it is seen), as there was no space elsewhere for the complete track. The actual sound of the car is to be heard coming into the picture and going away again; therefore the 'heads and tails' (handles) need to be added to complete the effect. In order to synchronize this accurately it is not necessary to have relevant time codes, only the relevant sound with synchronous code for running in lock into the workstation. This sound is transferred into the workstation and labelled . It is now held synchronously in the workstation. The original and the edited version of the sound can now be synchronized; the 'complete' sound is now placed on a track on the time line next to the shorter edited version. A rough first guess of the synchronization is made and the two are locked together. Next, both tracks are heard at once, and the original is then trimmed against the edited master. The two sounds will then merge as one, first with a slight echo, and then a phasing effect is heard – which merges into one sound and synchronization. Alternatively, the sounds could be displayed as waveforms graphically on the time line and matched up sound wave to sound wave. A final check is made that the sound is correct and the track is laid for the sequence.

Next, a sound effect already chosen and recorded into the library is laid. It is not suitable so another is chosen from a CD sound effects disc; this time it will be viewed and recorded onto the appropriate track. The workstation with video following is started. Track 1 on the workstation is selected for the recording. The CD output is routed to the track, track recording begins and the sound is therefore recorded through the shot. The beginning of the programme is now found and the track discarded to the start time code point; the track is played and at the end of the scene at the time line the track is cut away. The track has been automatically arbitrarily labelled by the workstation; the operator now renames the track and moves it into the correct directory. The next effect is chosen and laid on track 2; it is suitable. The third sound effect, now at a different location in the countryside, is moved to the time line and played with the picture. Unfortunately, it runs out before the sequence finishes. However, since the effect still exists in the library it can be used again and it is joined to the previous section of the same effect. To make the join imperceptible, the crossfade facility is used between the two cuts and the effect is just long enough. This section could be looped using the same fade parameters on each join if needed.

To lay the track of the street vendor, the studio is used. The time code of the start of the vendor is noted and a colleague shouts the words as the time code is reached. The workstation is set to loop and insert record across the in and out points of the three words. Recordings are stacked up as 'virtual tracks'; the last is the best so it is kept on the track and labelled and stored in the library. The recording is moved to track 2.

Since only two tracks were available on the original edit, it is now necessary to move, or split, some of the tracks to make mixing easier. This often happens when there is limited track space and tracks have to be butted together. More often than not, it is dialogue that needs to be split into two or more tracks as there may be sections with background sound levels that change or 'step' at each change of shot. In the series of interviews taken at the busy street corner, none of the shots have the same level of background traffic noise. As a result, the incoming track, at the end of an edit, is likely to 'step' in level to the next edit. This discontinuity in sound may be disturbing to the viewer. To overcome this, a further soundtrack is added, which is a wild track of the background sound recorded at the location during the shooting. This is copied from the master video using a note left by the picture editor; it was in fact transferred into the workstation's library with other sound effects after the spotting session. To help the mixing process further, the dialogue track is not left butted together but is split apart and laid alternately ('checkerboarded') on to a separate track so that the sound mixer can adjust the levels of the tracks, by fading down the first track and slowly fading in the second alternating track, thus creating an imperceptible join. Ideally the handles of these shots would have been added to smooth the mix but unfortunately the dialogue was spoken far too quickly to make it possible to replace the handles of each of the separate takes. This would be somewhat of an ideal, for interviews never seem to cut where pauses allow overlaps to be added. However, this technique can be used in drama where lines of dialogue should begin and end in silence.

When splitting an edited track into separate tracks, the first operation is to move the picture to the cut point, the audio following precisely in lock. Next the system is told via keyboard or cursor to cut the sound on track 1 at the assigned point on the right-hand side of the time line, the 'other' sound is cut to track 2 and pasted on. In this way a butted dialogue track can be off-laid onto the various tracks of the system. The cueing is accurate to better than one frame, and any point can be accurately found. Using similar techniques it is possible to fit wild recordings of dialogue into close synchronization. If it is necessary to go back to the original sound masters which are held within the library, these can be selected from the library page. The access to this is almost instant.

Although sounds can be recorded and moved around the cue sheet, none of the sounds from the library are erased so they can be used as many times as is required. Workstations can also make up loops of sound with perfect joins. Some even offer piano-type keyboards to lay spot sound effects such as foot-steps and clothes rustles. The soundtracks are now ready to be mixed, either by being played off the workstation or perhaps taken to another mixing suite. In this situation the electronic cue sheets will be lost if, for example, the tracks are transferred to an eight-track modular digital multitrack dubber. Here the cue sheets can be produced manually. These will specify the points at which each sound starts and finishes, together with information as to whether audio is faded in or out, mixed with another track or simply left out or cut. It may give stereo positioning information as well. Some recording consoles can also produce graphical displays once the sounds have been loaded in. These charts are also useful in simple linear editing suites when mixing audio.

Networks

General purpose data networks are regularly used in computer systems.

Networks are also used in digital audio for transferring files from one system to another, perhaps within a studio complex – a local area network (LAN). Here audio data can be transferred from one workstation to another to allow sound mixing rather than editing to be carried out in a more suitable suite, perhaps where a voice-over studio is available. A server could also provide a central library of sound effects available on a dedicated workstation and accessible by everybody on the network. The device providing the infor-mation is the network server, the device receiving the information the client. In audio post-production, to be really useful the network needs to be able to work at speeds greater than real time and handle a number of files at once.

Advantages and disadvantages of workstations

The advantages of work stations are:

- The systems are very inexpensive when used with a standard host com-puter
- Sound remains in the digital domain and is therefore of the highest quality.

- Access is instant (although this will depend on how full the system is).
- The system can shuttle in synchronization at slow speeds and park in synchronization if the video recorder is an integral part of the workstation. Otherwise it may depend on the stationary accuracy of the interlocked video machine.
- Improvements and updates can be made easy to incorporate into the system by upgrading software.
- It is possible to diagnose faults via a telephone line through a 'modem' to increase maintenance efficiency. (The system is directly interfaced to the telephone line.)

The disadvantages of workstations are:

- The system can be expensive in dedicated computer form.
- The capacity can be limited and depends on price.
- Work has to be loaded into the system.
- It may not be possible to remove the stored data discs physically from the system, making it necessary to copy 'archive' material out of the workstation at real speed.
- The systems are not operationally compatible.
- There are dangers of the system crashing and losing all the completed work within the system.
- It is highly complex technology.

12

Stereo

Stereo sound is found in most domestic environments. CDs and the radio are in stereo, and even archive material tends to be simulated stereo. Television may or may not be, and in the cinema we find multi-channel stereo surround sound. Surround sound can be defined as any stereo system that uses at least left, centre and right channels across the screen with one or more surround channels. This is available in the home, as home cinema, using video cassettes, laser discs, DVD discs, television transmissions and computer games. One format, the Dolby Surround format, is a fully stereo compatible format which means that most feature films transmitted on television are automatically in this format.

Stereophonic sound conveys information about the location of sound sources, but this ability to provide directional information is only one of the advantages of stereophony. Stereo recordings have an improvement in realism and clarity over mono recordings. They have a clearer distinction between the direct and reflected sounds which produces a spacious three-dimensional image with a better ambient field. Pictures can be powerfully reinforced by the use of stereo sound with better directional cues for off-stage action, and with sound effects which have increased depth and definition.

Even television, with its small sound field, benefits from stereo sound. The positioning of sounds may not be practical but the re-creation of an acoustic can be used advantageously. In a multi-camera production it might be impractical to position the microphone correctly for each shot in a rapidly cutting sequence; however, a general stereo acoustic can give an impressive result, particularly in sports and current affairs programmes using surround sound.

Recording stereo sound and surround sound requires more equipment and more tracks. Post-production time is increased and there are small increases in recording time on location or on the set. Thus stereo productions are more expensive to produce than those made in mono, in terms of both time and equipment needed. It has been estimated that it takes 25 per cent more time to post-produce in surround sound than mono.

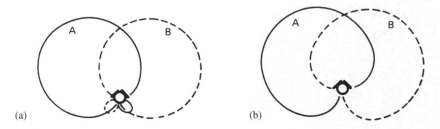

Figure 12.1 Polar diagrams for polar microphones used as a stereo pair. (a) A crossed pair of cottage loaf polar diagram microphones; (b) a crossed pair of cardioid polar diagram microphones

Stereo dialogue

AB coincident pair technique

It is said that the best stereo sound is produced using a pair of microphone capsules with cardioid response separated by about 6 inches at an angle of 110°. The sound arrives at capsule A at a different time to capsule B. There is a phase shift difference between the sound arriving at the two microphones, depending on frequency (which will also depend on direction) and the acoustics of the environment. It is this time and phase difference that produces the stereo effect.

This AB coincident pair technique is not suitable for television and film production dialogue recording – it is too critical to set up and takes time to arrange successfully. Using this type of microphone technique produces dramatic changes in the stereo image when a microphone is racked in and out on a boom across a set; the effect is inconsistent with the picture and may not be compatible in mono.

Spaced microphone technique

The spaced microphones technique is sometimes used to record stereo for television and film. The microphones cover the sound field at a distance from each other and this can produce excellent effects recording, particularly of moving objects. This technique is advisable for motion picture stereo matrix systems, where two spaced microphones help to reduce the tendency of phase incompatibility in mono.

It can be used for dialogue recording, but needs even more careful setting up than the A/B coincident format.

M & S microphone technique

It is possible to replace the two cardioid microphones in the coincident pair with two figure of eight microphones and again produce effective stereo; but this again is unsatisfactory from an operational point of view.

However, by using a figure of eight microphone and a cardioid microphone and then reversing the phase of one of the microphones and moving it 45 degrees to the other, a very effective 'cardioid' stereo pickup is available

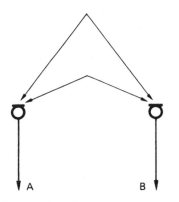

Figure 12.2 A spaced pair of stereo microphones

(Figure 12.3). One microphone will pick up the sound 'in the middle' and the other pick up sound 'sideways'. The output from this stereo combination is called M and S respectively – Mid and Side. Tradition has it that the positive north lobe or pickup pattern of the M is kept in phase with the positive S lobe of the side. The angle between these lobes can be varied by altering the proportion of the side signal to 'centre', thus offering control over the width of the image. A matrix is required to decode the recording into a standard left plus right system. The system has several advantages:

- Since the mono signal is produced from a single microphone, the problems of mono incompatibility are fewer.
- The width of the stereo image can be varied to match the picture by the ratio of the M to S signal.
- The control of image width can help match sound, shot to shot.
- The mono signal can be monitored and recorded in the normal manner with the side signal recorded on track 2, which can be ignored except for the occasional checking.
- Phase errors in the system produce only a change in width, which is less disturbing than the 'image-wobble' which other systems possess. In the

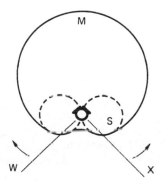

Figure 12.3 The response of a figure of eight and a cardioid microphone used in the mid plus side format. A special matrix is needed to decode the signal to a left and right format

case of complete phase reversal, one channel only will result, the image being changed from left to right. If there are phase and level problems in an M & S recording, the result is a loss of separation. (On portable tape recorders the small size of the meters does not allow easy, accurate line-up, and extreme care must be taken when setting up recording levels for M &S recording.)

M & S is an excellent microphone technique to use in recording dialogue if little or no post-production is to take place. It produces consistent results and can always be reverted to mono if the stereo produced is unacceptable. Unfortunately, the signals have to be decoded into normal left and right stereo. This means that a matrix has to be used for monitoring on location and in the audio post-production suite.

There are further problems if a microphone boom is used. In mono it is normal practice to favour the artiste speaking by panning the microphone head. If this is attempted with a stereo microphone, the stereo sound field will be heard to move, although the picture will remain static; a compromise position has to be found. Therefore, in mono reproduction some artistes are 'off mic' and subject to increased background noise – this is a general problem with stereo. Unfortunately, this noise also includes more concentrated sounds as well as ambient stereo sounds, such as camera noise, tracking noise, traffic, etc. In addition, all stereo recordings can suffer from some phase problems which mono will not. Some would therefore argue that where audio post-production is involved, properly recorded mono dialogue is of more value than stereo dialogue, provided that good stereo sound effects and music are available.

In the motion picture industry, dialogue is always recorded in mono and occasionally it is moved from the centre of the screen to the sides for dramatic effect. In fact the matrix used to record four-channel stereo in motion picture work is in some respects phase dependent. Thus, the dialogue recorded in stereo may be unsatisfactorily reproduced (phase similarities tend to narrow the stereo image). The motion picture industry tends to create stereo by careful track laying and mixing, rather than by original stereo dialogue and effects recording. Music is always recorded in stereo, possibly with additional channels that can be added to create a special effect during the mix. The spots effects are often in mono, as are other single source sounds, which are moved to position across the sound field by panning, but atmospheres and other multi-source sound effects may well be recorded in stereo.

Multi-channel stereo sound systems

For simple video stereophonic productions the television industry records much of its original sound in stereo.

Stereo exists in a universal simple two-channel form; however, surround sound has many differing formats. In the late 1980s an agreement was made in the USA that the minimum number of channels for digital film sound was to be five – called 5 point 1 channel sound (the 'point 1' refers to a separate low-frequency channel). Surround sound systems range from four to seven channels:

● Dolby Stereo/Surround system: four-channel analogue encoding system recording onto two tracks – screen left, centre and right with audience surround channel; it is fully stereo compatible and is used in films, television, laser discs, video cassettes, radio progammes, CDs and interactive computer games.

- Dolby Digital is a five-channel digital system – screen left, centre and right with left and right surround channels, used in films, laser discs, digital TV systems, DVD (digital versatile discs). Recorded directly onto the film between the sprockets.
- SDDS (Sony Digital Dynamic System) is a seven-channel digital sound film system recorded directly onto the film.
- DTS (Digital Theater System) is a five-channel digital system using CD-ROM discs, synchronized via time code optically recorded on the film. It produces the highest technical quality of the digital sound track formats and is available in consumer formats as well such as DVD.

Surround Sound from Dolby is available to audio post-production suites as software interfacing with specific digital audio workstations, allowing the four recording channels of screen left, right and centre plus surround to be matrixed into standard Dolby-compatible two-channel stereo. The software provided keeps the entire system in the computer's digital domain, and a system is provided for calibrating the acoustic levels within the post-production suite itself. Sounds can be moved around the four-channel sound field with a mouse, a screen grid showing the position of the cursor representing the more usual joystick. (This means a joystick is unnecessary within the recording console.) Like any matrix system, the two encoded tracks can suffer from some interference problems. Centre information, because of the system, will be unavoidably recorded in the right and left decoded channels and information panned left or right will appear in the surround channel, although the results are more than acceptable. It is as well to check stereo compatibility at an early stage to see if there are to be problems. This applies to all multi-channel formats that may be played in mono where no specific mono mix is being produced.

Track laying and stereo

Stereo soundtracks are track laid and mixed in a similar way to mono ones, but there is the additional advantage of being able to position the sound in the sound field during the recording and mixing process. To ease re-recording problems, it is important that the same stereo recording format is used within the workstation itself, and it is sensible to place stereo sound correctly in the pre-mix as well. It is vital that an M & S system is not butt joined to an incompatible A/B stereo recording. It is also important to check that moving sounds match their moving pictures. A train, for example, should be laid moving left to right – if the train is shown in the picture moving left to right (although its direction can be changed in the mix)!

Stereo recordings should, wherever possible, be made on tracks that are within the same machine rather than tracks that are physically tied by time code synchronization to another. This will reduce any possibility of phase errors occurring between the sources.

One of the major problems with stereo is that it increases the number of tracks needed since, of course, each stereo sound requires two tracks. Within the workstation systems the number of tracks available is limited, although time code synchronization allows an unlimited number of machines to be locked together.

When laying soundtracks within non-linear picture workstations offering only two audio tracks, there is almost no possibility of using stereo recordings; each stereo soundtrack creates problems since tracks can only be butted. No longer is it possible to allow sounds to extend over cuts, with split edits(L cuts), and nothing can be laid under another sound. A two-track machine now only has available one soundtrack for the picture. To extend the number of soundtracks available, it is necessary:

- to lay sound to picture in mono using the tracks only for guidance, the edit decision list providing information to lay back the stereo sound from the stereo source masters;
- when EDL facilities are not available, to lay sound in mono and make careful notes regarding source tapes for the final track laying;
- to upgrade the non-linear picture editor with more tracks for particular projects;
- to lock a separate sound machine to the workstation while editing, specifically for sound work.

Stereo audio synthesizers

It is sometimes necessary to re-create stereo from mono, either in audio postproduction or at a transmitter, in an attempt to maintain a 'stereo' output even when mono is being reproduced. Devices to produce this 'pseudo stereo' sound have been available for a number of years.

The earliest attempts at producing a stereo image from a mono sound source were made in the late 1940s in Germany; the system developed used a high-pass filter on one channel and a low-pass filter on the other. This gave the illusion of the bass instruments of an orchestra coming from one side of the sound field and the treble instruments coming from another.

This was developed into a more sophisticated system of two interleaved comb filters, one for the left and one for the right. The combs fall at random intervals, and if they are heard together in mono, no audio coloration is noticeable. Instead of comb filters it is possible to use frequency band filters. But in this case it is difficult to construct interleaving frequency bands that do not add coloration to the sound when reproduced in mono.

Time delays are also used to produce pseudo stereo, but if the delay times are not carefully controlled mono incompatibility will result.

Stereo synthesizers are designed to process both mono dialogue and music into stereo, but they are far more effective with music. Speech tends to be monotonic, with a disembodied quality, and this means that the depth and width controls provided cannot be used to their fullest extent. The exception is dialogue recorded in a very reverberant acoustic, where full width and depth control increases intelligibility and appears to reduce the background reverberation.

13

Music

Music has always been a vital part of the film and television soundtrack. It is used to create atmosphere, mood or emotion, and appears in almost all types of productions, whether they be motion picture films for theatrical release or simple corporate productions, which merely have title music. Title music will provide the mood of a production, while background music will provide the atmosphere and drama. More recently, music video promotions have used the techniques of audio post-production to produce special music mixes with synchronized sound effects.

Music is laid in exactly the same way as soundtracks. Just as certain tracks are designated for sound effects, for example, so certain soundtracks will be designated for music. Similarly, music can be specially composed and recorded or more likely can be ordered from specialist libraries.

Use of music

Whatever type of music is used, it must be placed at the right point for the right reasons. The most frequent failure in music editing is placing music in the wrong place, for the wrong purposes. It is important to recognize where the principal impact of music should be in a particular scene. Where does it come in? When does the music go out? Perhaps it can be sneaked in unobtrusively, using a sound effect to mask the entrance. Alternatively, it might be introduced on a dialogue cue in a quiet, reflective or even thunderous mood. Entrances can be understated, overstated, or just stated.

Action and sound effects are good counterpoints to music and can be highly dramatic, expressing the emotions of characters rather than merely accompanying them with the speed of the action. However, over-impressive sounds can detract from the pictures. Loud music is also undesirable since, unless it is carefully composed and positioned, it will need to be held down low under other sounds. It is important to avoid excessive high-frequency content in music too, since if this is held down in the mixing stage, the music will lose its body and characteristics.

It is essential that music does not fight with dialogue or effects. Small ensembles of strings or woodwind, for example, work well behind dialogue providing they are in careful register and timbre, but a significant change in level does not usually work unless the music is punctuating the dialogue to produce an effect. Perhaps the most startling effect is sudden silence. Halting the music and playing the moment of drama in silence can be intensely effective.

Music especially composed for a production

If music is to be especially written for a production, the composer and director will first decide on the sections that are to have music written over them, spotting. The composer will be given a video copy of the edited material, in as near a fine-cut version as possible, with burnt-in time code (a window dub) and possibly a track, giving the proposed tempo. This may be recordings of a piece of music similar in feeling to the type required – a temp track.

Until the introduction of the domestic video cassette recorder, composers only had a limited time to view their films or programmes and to prepare notes for composition. Since the introduction of the domestic VCR, composers have been able to view whenever they wish and construct tracks using Midi computers at home, and then take the time coded videotape and computer information to the studio for the final recording and mixing. Using digital audio workstations, the final music track may well be composed and recorded entirely in the home studio. This option is highly attractive since it offers a greater opportunity than ever for low-budget productions to have their own specially produced music.

Music in audio post-production is classically divided into two distinct types: source music, and music scored for dramatic effect.

Source music comes from an actual source on the screen. This may be visual or just implied. People dancing to a 'live' band is scored with 'visual' source music, but if the band is playing on the radio, it is 'implied' source music. This music can be developed into part of the framework of a film and it may then become the score of the production. Often the theme is set up in the main title and from then on is used wherever appropriate.

Dramatic scoring is the most usual form of musical scoring. Basically, here the music is being used in a theatrical way – the composer has complete freedom to decide upon how the music is written and how he or she portrays the emotions of the production.

Once the composer's ideas have been approved, and the final cut has been accepted or 'locked', the music recording session will be scheduled. This is usually the last operation before the final soundtrack mix; the music may or may not be recorded with the picture.

Music recording to picture

In this method, the completed final cut of the scene to be recorded is viewed by the conductor and the musicians as they perform the work. The film or the videotape is marked with cues to indicate to the conductor when the music is about to begin and end. Through headphones the conductor, and possibly the musicians, will hear the click track, or an electronic metronome dictating the tempo or the exact required beat of the music (tempos are expressed in frames rather than in metronome beats). Cued from time code or film footage readings, the click track runs in synchronization with the picture. It can be started at a number of clicks prior to the first frame of picture so as to give a count-in to the musicians; the click generator can be programmed to speed up or slow down at the appropriate point with the picture.

When synthesizers are a major source of music in a production, the picture time code is locked to the synthesizer's Midi code. This allows a number of keyboards, electronic drum machines and other Midi devices to be accurately controlled to the picture time code. Computers of this type can also calculate tempos, real and relative timings, set accelerandos and retards to hit cues at the correct time, and do all the paperwork necessary for music recording to picture. All these systems have been developed for the home studio as well as the music recording studio.

The ease of synchronizing video recorders and digital audio workstations has led many composers with small studios to offer music/picture recording facilities to the audio post-production industry. These studios specialize in recording music that is to be reproduced in the home environment; this is ideal for stereo television work. However, recordings made for the domestic environment may not be suitable for reproduction in a cinema.

Music for motion pictures

Stereo mixes for the cinema need to be monitored on at least the three main channels, left, centre and right. This means that there is incompatibility with stereo music mixes that have been specifically recorded for two-channel systems – such as for the home. These tend to sound too narrow when played in a theatrical environment, because the coding matrix used in the cinema tends to pull the sides of a stereo image to the centre. This problem can be eliminated by monitoring through a matrix unit, allowing multiple speakers to be used during the recording session (simulating a matrix recording). Ideally, these multitrack tapes should be mixed together in a film mixing stage where the acoustic characteristics are correct for reproducing in a cinema, and the monitoring level is set to the recommended standard.

Music recording for the cinema is shown in Figure 13.1.

Often it is impossible to record music to picture. Perhaps the recordings have to be completed before the final edit, or perhaps there is insufficient finance for an expensive picture recording session. In this situation the music is recorded 'wild', and eventually a certain amount of music editing will be necessary to fit the music to the pictures. If the music is to be part of a series, 'atmospheres' that can be used at the appropriate times will be composed. Obviously, title music and credit music, together with breaks, are needed, and in addition to this, chase, comedy, fight and dramatic stings, for example, are recorded, all to be placed in the appropriate positions by the editor.

Very often music cannot be specially recorded for a production because of budget constraints and in this case mood music libraries provide a suitable alternative. Most mood music publishers issue a detailed catalogue listing the compact discs they are able to supply. The selection is enormous. Almost every type of instrument, orchestra or national sound can be supplied, to any

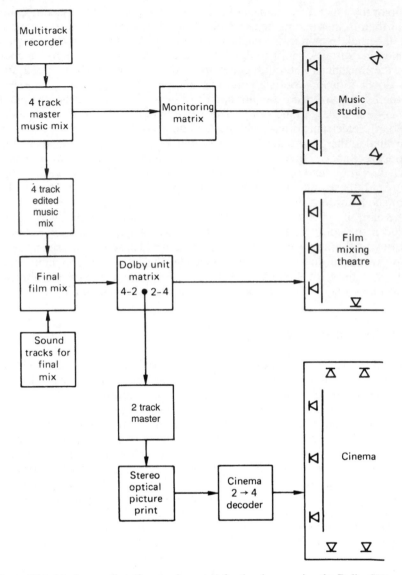

Figure 13.1 Music recording (four-track master) for the cinema using the Dolby SVA matrix system

mood, whether it is lighthearted or dramatic. These 'non-commercial' libraries are being continually updated and follow fashion well. A music library of specialist stings and links is shown in Figure 13.2.

Complicated editing and animation sequences requiring music are usually edited to already recorded compositions. This allows the picture editor the opportunity to illustrate the music and exploit the musical rhythm as he or she wishes. These techniques are always used in animation where musical tempo is used to structure the sequence.

CARLIN

SHORTS/LINKS/STINGS 2

SPORT/ACTIVITY
1.1 RESULTS ROUND UP 0'08
2.1 ACTION BREAK (A)-(E) 0'05
3.1 SPORTS DESK (A)-(C) 0'15
4.1 SPORTS FILLER (A)-(B) 0'10
5.1 OPEN ROAD 0'34

NEWS/FANFARES/LOGOS/IDENTS
6.1 NEWS ITEM (A)-(B) 0'04
7.1 BREAK (A)-(B) 0'07
8.1 HORSEPOWER 0'11
9.1 NEWS FILLER (A)-(B) 0'06
10.1 NEWSFRONT (A)-(B) 0'08
11.1 RACEWORLD (A)-(B) 0'08
12.1 PUSH BUTTON (A)-(B) 0'08
13.1 FANFARIAL 0'10
14.1 FANFARE (A)-(B) 0'13
15.1 FANFARE FOR GRAND ENTRANCE 0'11
16.1 CAESAR'S ENTRY (A)-(B) 0'13
17.1 LIGHT LOGO (A)-(G) 0'06
18.1 BRIDES ENTRANCE (A)-(B) 0'18
19.1 XMAS JINGLE 0'10

ROCK
20.1 HEAVY RIFF (A)-(H) 0'13
21.1 ROCK STATEMENT (A)-(E) 0'09
22.1 HEAVY BLUES 0'07
23.1 ROCK RHYTHM & BLUES (A)-(J) 0'13
24.1 STEPPIN' OUT 0'42
25.1 HILLBILLY ROCK 0'57
26.1 BLUES BROTHERS 0'24
27.1 SCRATCH (A)-(B) 0'22
28.1 SHIPPING UNITS 0'32
29.1 CHART STING 0'04
30.1 ROCK CELL 0'45
31.1 DISCO RAP 0'15
32.1 CHART STAB 0'03
33.1 CHART RUN DOWN 0'32
34.1 BLUES GUITAR 0'48
35.1 POPSHOT (A)-(N) 0'03

PERCUSSIVE
35.1 MODERN MACHINE (A)-(B) 0'40

TRAVEL
37.1 EXOTIC STING 0'08
38.1 BESIDE THE SEASIDE 0'05
39.1 CHINESE LOGO 0'06
40.1 BRIGHT EYES 0'33

PASTORAL/NATURE
41.1 ROMANTIC CHORD 0'08
42.1 WATERY (A)-(D) 0'35
43.1 WONDERMENT CHORD 0'29
44.1 CASCADING WATER (A)-(C) 0'27
45.1 WATERLILLY 0'06
46.1 FALLING LEAF 0'09
47.1 SIGHS 0'35

40.1 NEWMORN (A)-(C) 0'56
49.1 RAY OF LIGHT (A)-(D) 0'30
50.1 SPARKLE 0'03
51.1 WARM WONDERMENT 0'22
52.1 RABBIT 0'41
53.1 BUSYBODY 0'39
54.1 FLEDGLING 0'57
55.1 BREATHY TRILL 0'03
56.1 FLUTE SCENE (A)-(C) 0'09
57.1 BIRD IN FLIGHT 0'50
58.1 SLEEPY BIRD 0'55
59.1 FLYING HIGH 0'23
60.1 AFRO BONGO 0'06
61.1 HARP CUE (A)-(B) 0'05

EARLY INSTRUMENTS
62.1 AGINCOURT 0'35
63.1 COUNTRY FAYRE 0'36
64.1 COURTLY LOVE (A)-(B) 0'56

PERIOD
65.1 ROYAL BANQUET 0'54
00.1 SUMMER IS ICUMEN IN 1'00
67.1 THE ROAST BEEF OF OLD ENGLAND 0'24
68.1 THE QUEEN'S GARDEN 0'24
69.1 CHANCELLOR'S ENTRY 0'22
70.1 CHILDREN'S DANCE 0'26
71.1 BAROQUE STYLE 0'41
72.1 TUDOR ROSE 0'20
73.1 PRINCE REGENT 0'14
74.1 SYNTH BAROQUE 1'06
75.1 ENTERTAINER (A)-(B) 0'08

ROMANCE
76.1 HAPPY IN LOVE 0'22
77.1 SOFT GUITAR CHORD 0'06
78.1 THOUGHTS (A)-(G) 0'09
79.1 ROMANTIC THOUGHTS (A)-(E) 0'00
80.1 ROMANIIC EXPECTATIONS (A)-(B) 0'29
81.1 ROMANTIC EXCITEMENT 0'07
82.1 ROMANTIC DAWN (A)-(B) 0'34

SEA SHANTIES
83.1 THE GIRL LEFT BEHIND ME 0'07
84.1 PORTSMOUTH 0'07
85.1 BLOW THE MAN DOWN 0'12
86.1 BOBBY SHAFTOE 0'07
87.1 THE DRUNKEN SAILOR 0'06

SPACE
88.1 METEOR 0'05
89.1 SPACE CHORD 0'38
90.1 TARDIS 0'09
90.2 LOW DRONE (A)-(B) 1'12
90.4 SPACE PANIC 0'41
90.5 DEEP SPACE 0'35
90.6 UFO 0'34
90.7 AMAZEMENT CHORD (A)-(B) 0'45

91.1 DARK HOLE 0'40
92.1 TRAJECTORY (A)-(C) 0'20
93.1 ABANDONED SPACESHIP 0'40
93.2 SPACE SOUND 0'10
94.1 SPACESHIP 0'07
95.1 BEYOND THE PLANET 0'22
95.2 PASSING MOON 0'14
96.1 VOICE ARPEGGIO (A)-(C) 0'16
97.1 SPACE SUNRISE 0'19
97.2 SPACE SUNSET 0'19
98.1 COMET 0'05
98.1 VOYAGE INTO SPACE 0'50

ELECTRONIC KEYBOARD EFFECTS
99.1 WOBBLE BOARD 0'03
99.2 HIGH WOBBLE 0'04
99.3 WOBBLE STROLL 0'07
99.4 WOBBLE BOUNCE 0'07
99.5 HELICOPTER 0'10
99.6 SHORTALARM 0'02
99.7 HEAVY BELL WALK 0'16
99.8 MAGIC TRICK 0'07
99.9 GIANTS WALK 0'10
99.10 MACHINE SHOP 0'05
99.11 STEAM TRAIN 0'07
99.12 STEALTHY PLOD 0'14
99.13 CAR HORN TUNE 0'03
99.14 NIGHTMARE ORGAN (A) 0'18
99.15 NIGHTMARE ORGAN (B) 0'08
99.16 CREEPING 0'09
99.17 HIGHER AND HIGHER 0'05
99.18 LITTLE TINKLE 0'10
99.19 DEEP CHUFFER 0'12
99.20 RATTLE LEG 0'12
99.21 CREEPING UP 0'07
99.22 TINKERBELL 0'15
99.23 SNOWDROPLETS 0'08
99.24 PIANO ALARM 0'18
99.25 COLD WIND 0'06
99.26 BI-PLANE AWAY 0'20
99.27 DESCENDING KEYBOARD (A)-(B) 0'13
99.29 KNOCKING (A) 0'03
99.30 KNOCKING (B) 0'10
99.31 BUBBLE TIME 0'07
99.32 QUICKLY 0'03
99.33 MR ROBOT 0'07
99.34 TAPPING (A) 0'02
99.35 TAPPING (B) 0'10
99.36 RESONANT 0'09
99.37 FALLING SCALE 0'09
99.38 FALLING CHIMES 0'09
99.39 SWIRL 0'15
99.40 TWIDDLE 0'02
99.41 SLOW WIND 0'26
99.42 DEEP PERCUSSION 0'07
99.43 KNOCKING ECHO 0'15

See booklet for full details and composer credits   Cover design by Ian

CARLIN PRODUCTION MUSIC
Iron Bridge House, Bridge Approach, London NW1 8BD
Tel: 071-734 3251 Fax: 071-439 2391

Figure 13.2 A music library of specialist stings and links (*Courtesy of Carlin Production Music*)

Musical programmes

Recordings of live performances of musical works have to be carefully mixed so that they appear to match the pictures, although it is often impossible for the sound to slavishly follow the shots. In a reverse angle, for example from the rear of the stage, the stereo image would need to be reversed to be 'realistic'. Nevertheless, it is important that what is clearly seen is clearly heard. The ambience of the sound must fit the apparent acoustics of the location. In a concert hall, for example, the sound heard from the auditorium will often appear to be too distant for close-up pictures, so a closer perspective is required. Certain specific effects may be positioned to match pictures, and this technique is often used in rock music shows.

Copyright

All creative works are held in copyright by their owners, whether they be sound effects, recordings, scripts, music, or whatever. To make use of such works, the user is liable to pay a fee. Usually the fee allows limited use of the work rather than an outright sale, although contracts for outright sale are not usual for the purchase of sound effects and some music libraries.

Figure 13.3 shows a disc from a 'copyright assigned' library.

Sound effects are licensed directly by their owners but scripts may be licensed out by agents as well as writers. Music rights are controlled by specific organizations set up, particularly, to deal with contracts and copyright. They take vigorous action in defence of their clients, who often give them complete control over copyright collection.

Copyright law is complex and varies from country to country but there are reciprocal agreements to allow copyright to be administered across national boundaries. Copyright clearance should be obtained prior to using any work. In recorded music, clearance often needs to be obtained from the composer, the owner of the recording of the music, the owner of copyright held in the performing of the music, and the owner of copyright held in the broadcasting of the piece.

The fees payable will depend on how large the audience for the proposed

1	*Victorian* . . . .*(4m.47s)*
2	*Edwardian* . . . .*(4m.58s)*
3	*1920's* . . . .*(4m.51s)*
4	*1930's* . . . .*(4m.52s)*
5	*1940's* . . . .*(4m.54s)*
6	*1950's* . . . .*(4m.57s)*
7	*1960's* . . . .*(4m.58s)*
8	*1970's* . . . .*(4m.59s)*

The purchaser of this Cd is licensed to duplicate the music herein contained for the purposes of video, audio visual, cinematographical or multi-media productions. This right is strictly non-transferable.

COMPACT
disc
DIGITAL AUDIO

Figure 13.3 A disc from a 'copyright assigned' library offering full buy-out rights (*Courtesy of Trackline Music, Cheshire*)

production is. Will it be shown only for industrial use? This would incur a small fee. Alternatively, will it be for worldwide television distribution? This would require a large fee.

Organizations concerned with music clearance include:

- The American Society of Composers Authors and Publishers (ASCAP)
- Broadcast Music Incorporated (BMI)
- The Harry Fox Agency (USA)
- Mechanical Copyright Protection Socicty (MCPS)
- Performing Rights Society (PRS)
- Phonographic Performances Limited (PPL)
- Society of European State Authors and Composers (SESAC)

14

Monitoring and the environment

Soundtracks are reproduced in various different acoustic environments: the home, the cinema, the conference hall. The mark of a well-mixed soundtrack is one that can be heard intelligibly in all these different situations. In an ideal world, all control rooms and listening rooms would be standardized, to allow perfect recordings to be heard. However, they are not, and since sound is very subjective, and since no two people are likely to interpret sound in the same way, it is surprising that there is any similarity between recorded sounds at all.

The control room

A control room for monitoring sound must possess certain characteristics:

● The sound balance recorded in the room should not sound any different when played back in other environments.
● As a listener moves around the room, the sound quality should not change noticeably – although, of course, the stereo image will.
● When listening at normal listening levels, within a digital environment it shouldn't be possible to hear the background noise of the recording medium, unless there is a problem. This enables a check to be made if there is a build-up of noise through the system.

In the audio post-production control room, decisions are made during the sound mix on:

● artistic judgements of levels, perspective of sounds, fades, dissolves and cuts, special effects;
● correct synchronization, making sure there is the correct time relationship between spot effects, dialogue, music, etc.;
● technical quality concerning the acceptable frequency response, phase errors and noise and distortion of the system;
● placement of the stereo image, and surround sounds, making sure that the sound moves correctly in relationship to the pictures across the stereo field;

● stereo and surround sound compatibility with mono, making sure that viewers in mono receive an intelligible signal.

Figure 14.1 shows a small audio post-production complex.

Monitoring loudspeakers

The most important instrument in the audio post-production studio is the loudspeaker; from it, everything is judged. It has been said that loudspeakers are windows through which we view sound images, and this is a fairly apt analogy. However, sounds are more difficult than pictures to interpret correctly; an experienced eye can easily check the colour balance of a colour photograph by comparing it with the original. Sound, however, is open to a more subjective interpretation, being influenced by acoustics, reverberation, loudspeaker distance, and volume. All these affect the quality of sound we perceive through the window of our loudspeaker.

Loudspeakers are notoriously difficult to quantify. Terms such as harshness, definition, lightness, crispness and the more modern terms, translucence and sonic purity, are commonly used. But these different terms have different meanings for different people; stereo adds yet one more dimension to the problem.

For stereo monitoring, matched speakers are essential to ensure a consistent sound image across the monitoring window and acoustics must be consistent.

Stereo monitoring

Stereo is used in audio post-production in three main formats:

1 The simple left, right stereo format used in stereo television (two separate discrete channels).
2 The cinema left, right and centre with ambient format, used in Dolby optical stereo (SVA) (a matrix encoded system recorded on two optical tracks).
3 The five-track stereo surround formats used in film and the Digital Versatile Disc.

If stereo recordings are to be satisfactorily monitored, the sound image reproduced needs to have a fixed relationship to the picture. It is important, therefore, that there are sufficient sources across the sound field. In the cinema, where the width of the screen is perhaps 15 metres, at least three speakers are required to reproduce the sound satisfactorily, but in television, where the picture image is smaller, a pair of speakers positioned either side of the screen produces a satisfactory image.

It is generally recommended that in domestic two-channel hi-fi stereo sound systems, the speakers are placed at angles of 60° to the listener. Smaller angles make it difficult to judge the stability of the centre (or so-called phantom) image, and larger angles can remove the centre image almost com-

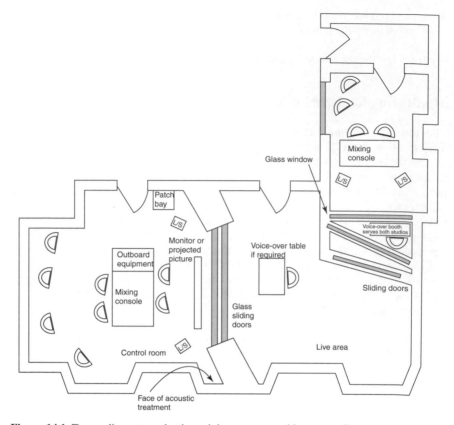

Figure 14.1 Two audio post-production mixing areas served by one studio area

pletely (leaving 'a hole in the middle'). At angles of 60° the listener and speakers are equally and ideally spaced from each other, the listener sitting at the corners of an equilateral triangle. This system has been adapted for television, where there is the additional requirement that the visual field and the sound must relate. Experience has shown that the width of the screen should be repeated between the edge of the screen and the loudspeaker beside it. This allows a conventional 60° pattern in a reasonably sized room. The room should be acoustically symmetrical from left to right, so that sound does not change as it moves from one speaker to another.

In a studio control room this set-up may not be ideal, for only one person may be able to sit in the correct listening position – a slight movement of the head leading to a significant change in the stereo image. This situation can be improved by far-field monitors with a larger sound field, allowing more people to hear the stereo image. This sound field is, however, more likely to be influenced by the acoustic characteristics of the room. If near-field monitoring is used, the loudspeaker systems need an exceptionally smooth frequency response over the acoustic area. When using Dolby Surround, set-up procedures are provided by the manufacturer.

Acoustics and reverberation

Reverberation affects the quality of reproduced sound and so it is a vital consideration in the design of any sound listening room, whether it be a studio or a cinema. Reverberation time is measured by the time taken for a sound pressure to drop by 60 dB; this can vary between 0.2 and 1 second depending on the size and the treatment of a studio. This reverberation time, however, must be the same at all frequencies to produce a good acoustic, and not just correct at some mid-reference point. It is not unusual for a room to have a short reverberation time (RT60) at high frequencies, but a longer one at low frequencies. This will, among other problems, produce an inaccurate stereo image. High-frequency reverberation can be treated with simple surface treatments. However, low-frequency problems often require changes in structural design.

After an area has been acoustically treated and the furnishings, equipment and speakers mounted, it is usual to take frequency response measurements in order to check the acoustics in the room. If problems do arise it is possible to equalize the monitoring system electrically, by modifying the frequency response of the speakers, to produce the desired acoustic response. However, this technique will not remedy, for example, deep notches in the frequency response, poor sound diffusion and poor stereo imaging. The electrical equalization of monitoring systems in small studios is therefore not necessarily an

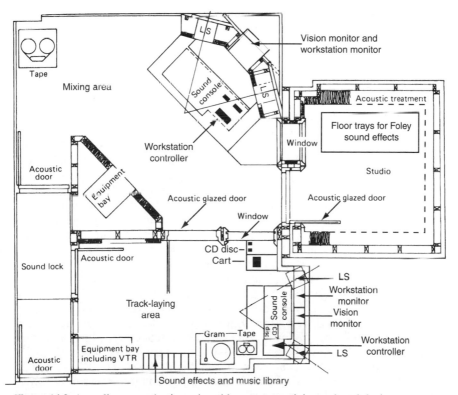

Figure 14.2 A small post-production suite with a separate mixing and track-laying area

answer to acoustic problems. It may well appear to produce an improvement at one point in a small room, but at other positions in the room problems may well be exaggerated. However, these techniques can be used satisfactorily in large environments such as cinemas.

Background noise

It is important that the audio post-production studio environment is quiet. Additional noise is distracting and can affect the ability of the sound mixer to hear accurately, although when mixing mono material it is reasonably easy to ignore background noise as the brain tends to eliminate sounds not originating in the monitoring speaker. With stereo and surround sound, however, background noise becomes more disturbing and it is difficult to ignore the distractions of ambient sound within a room (such as video players with cooling fans, buzzing power transformers, video monitors, air-conditioning, and passing traffic).

To assess how distracting external noise can be, architects use noise contour curves (NC curves). Measurements of the acoustic noise of a room are made, using special octave-wide filters. The NC rating is weighted to allow higher levels of noise at low frequencies, matching the sensitivity of the ear. A noise level of NC25 is a practical and acceptable level, making it just possible to hear fingers rubbing together with arms at the sides. (Figures between NC5 and NC10 are considered very good.) To achieve a suitable figure, noisy equipment should be removed from the monitoring environment.

The importance of listening levels

When mixing any type of programme material, it is important to monitor it at the correct volume. This will be determined by the level at which the material will eventually be reproduced. There is a strong temptation in the sound studio to monitor at high levels, so as to produce an apparently finer and more impressive sound. However, since the ear's frequency response is not flat, but varies considerably with listening levels, the temptation should be resisted. While very loud monitoring levels tend to assist in discovering technical problems and aid concentration, they will not assist in creating a good sound balance. It is all too easy to turn up the monitoring levels while being enthusiastic and enjoying a programme, forgetting the problems this will create. Once the sound level is set at the start of a mix, it should never be altered; it should be set to a normal reproducing level.

The requirements for film audio post-production monitoring rooms are set out in the International Standards Institute recommendations and the Society of Motion Picture and Television Engineers' publications, and are widely accepted. A sound pressure level of 85 dB has been adopted which is recommended to be 6 dB below the level at which an optical soundtrack will clip.

The introduction of digital soundtracks in films seems to have led to increased reproduction levels in some cinemas, no doubt because of the elimination of system background noise.

Computer sound

The growing number of CD-ROMs and computer games that are being audio sweetened in post-production suites, sometimes with multi-channel sound, adds yet another dimension to audio post-production. Quality from a CD-ROM can be of the highest but most games are recorded at a low-quality eight-bit rate, and, of course, material is replayed on small, very cheap computer speakers, rather than high-quality hi-fi systems.

Television monitoring conditions

The ANSI/SMPTE television control room recommendations are to be treated with more caution. They recommend a monitoring system with a response rolling of 1.5 dB per octave above 2 kHz. There is also a bass roll-off of 6 dB per octave below 100 Hz. This does not entirely reflect the frequency response of equipment in many high-quality domestic environments, particularly, of course, where one is trying to emulate a cinema environment using laser and DVD systems. The SMPTE figures are perhaps more suitable for worst-case checking.

Figure 14.4 shows the audio response of a good-quality stereo television receiver and a studio monitoring loudspeaker.

A television monitoring environment should try to simulate the listening conditions of the average living room. In Europe this size is likely to be about $5 \times 6 \times 2.5$ metres, with the reverberation time having little dependency on frequency (given average furnishings, it is about 0.5 seconds). Since the majority of television viewers live in cities where environmental noise is high and sound insulation poor, there is likely to be a high level of distracting noise. This is countered by turning up the volume, although consideration for neighbours will restrict this – giving a sound volume range available of about 35 dB in the worst case. An appropriate listening level in the home is likely to fall, at most, between 75 and 78 dB (compared to 85 dB plus in the motion picture theatre).

Television control rooms designed for stereo sound monitoring are likely to have:

- near-field sound monitoring with viewing by video monitor;
- speakers placed at three widths apart on either side of the screen;
- speakers subtending an angle of 60°;
- facilities to check mixes on high-quality and poor-quality speakers;
- a reverberation time of less than 0.5 s;
- the operator closer to the loudspeakers than the surrounding wall;
- a listening level set to about 75 dB SPL.

The introduction of wide-screen television has increased the effectiveness of stereo sound.

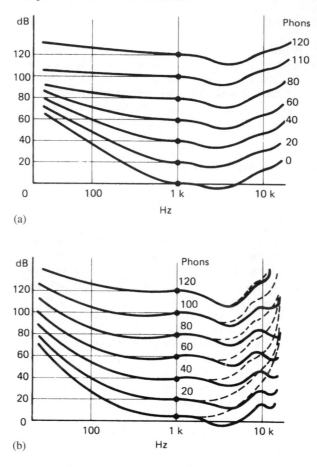

(a)

(b)

Figure 14.3 Equal loudness curves demonstrating that the ear is more sensitive to changes in volume at middle and high frequencies. (a) Fletcher Munsen curves; (b) Robinson and Dadson curves include the age of the listener: unbroken line indicates a 20-year-old, broken line a 60-year-old

Film theatre monitoring conditions

At least three loudspeakers are used to reproduce the stereo sound image across the motion picture cinema screen, designated left, centre and right. In addition there are surround channels within the theatre auditorium itself. In the digital formats five channels are used (5 point 1 sound), with two surround channels and low-frequency enhancement. In the Dolby Surround analogue format, four channels of sound are used, one being surround; all are encoded onto only two recording channels, via a matrix.

Through the matrix, the information on the left is fed directly to the left track. The information on the right is fed directly to the right track. The centre is fed at a reduced level to both in phase. The surround is fed at a reduced level to both, out of phase, and has a delay line in its output.

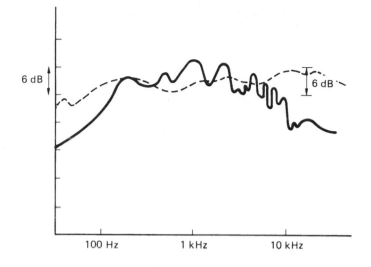

Figure 14.4 The audio response of (a) a good-quality stereo television receiver and (b) a studio monitoring loudspeaker

Matrixed systems are not crosstalk-free; this means that sound from the screen speakers can be heard in the surround speakers. However, this effect can be all but eliminated by adding a delay line to the surround speakers, and restricting the frequency response of the system to 7 kHz. The brain then identifies the first place from which a sound is heard as the sound's origin, and mentally ignores other sources of the same sound arriving a fraction of a second later. The signal from the front speaker reaches the ear first, so the mind ignores the same signal from the surround. (This is known as the Haas effect.) The position of the surround speakers is not critical, providing that there is a good balance between front and surround channels.

Multi-channel discrete systems have none of these problems, using, of course, discrete channels of digital audio sound.

Reproduction of motion picture film on television

More and more motion picture feature films are now being distributed for home use; indeed the majority of the income from feature film exploitation is derived from video sales. This means that it is in home environment, and probably in mono, that most productions are likely to be heard. Specifically mixed for reproduction in the cinema environment, these soundtracks may not be entirely suitable for domestic reproduction. In general, a cinema mix when replayed in the home will have the music and effects at too high a level. This means that the dialogue can become inaudible or unintelligible. In addition, dialogue equalization, used in cinema film soundtracks to improve articulation and intelligibility, can reproduce as sibilance on high-frequency transients, giving the impression of distortion in the domestic environment.

These problems are compounded in domestic stereo situations, for here the reproducing environment is flat and narrow with little reverberation, since the speakers are in a near-field environment. This is incompatible with the motion picture mixing theatre which offers a widespread image with more reverberation, and is a far-field recording environment. As such it is not particularly suitable for recording stereo sound for television. Any sounds of long duration tend to sound louder in a film theatre than they really are. Therefore, sounds that are powerful in a loud scene in a mixing theatre will sound somewhat less so at home. 'Mixing-up' is an answer. The stereo television sound is difficult to reproduce correctly in a motion picture mixing theatre, as the large distance between the two speakers means that there is, in effect, 'a hole in the middle' – which does not exist in the near-field environment.

Relative sound levels are shown in Figure 14.5.

So it can be seen that it is difficult to produce a good mix that will be entirely suitable for both the motion picture environment and the home environment. In an attempt to re-create the cinema experience in the home, surround decoders (Dolby Pro Logic, for example) are available for use with encoded video cassettes, laser discs and DVD discs.

Visual monitors

Loudspeakers in studios provide aural monitoring of the signal. To record these sounds accurately, it is necessary to have a precise form of visual metering. These two forms of monitor, the loudspeakers and the meter, must be carefully matched. The average recording level must correspond to a comfortable, recommended, relevant monitoring level within the studio environ-

Dynamic range, scale in decibels (these figures are very approximate)

130 db	Threshold of pain
	Sound level in clubs. Illegal in some countries
120 db	'Feeling' sound in the ears
110 db–115 db	Loudest sound on digital film soundtrack
90 db	Loudest sound on an analogue film soundtrack
80 db	Loud television set
70 db	Conversational dialogue
40 db–50 db	City background
35 db–40 db	City flat background
35 db	Blimped film camera at one metre
30 db	Quiet countryside background
25 db–30 db	Quiet background in modern cinema
25 db	Fingers rubbing together at arm's length
0 db	Threshold of hearing

Dynamic range digital sound track brackets 120 db through 35 db.

Dynamic range television transmissions brackets 90 db through 35 db.

Recording mixers shouldn't attempt to mix at continuously high levels as permanent damage to hearing may result

Figure 14.5 Relative sound levels

ment. Neither the monitoring volume nor the meter sensitivity should be changed if the sound level is to remain consistent. When mixing, it is usual for sound to be judged almost entirely from the monitoring systems, with only occasional reference to the metering, which provides a means of calibrating the ear. Two types of meters have been classically used for monitoring monophonic sound, and much is written about them in other books. They are found both on recording consoles and on the 'meter screens' of workstations.

The VU meter

VU meters are constructed in virtually the same way as AC volt meters and measure the mean average or RMS voltage of a sound signal.

For over 60 years the VU meter has been the American standard for visual monitoring. If the signal is intermittent, such as speech, the VU meter will indicate an average value. This will be considerably lower than the instantaneous maximum levels that are found in the material. (To compensate for this, engineers ride dialogue perhaps 3 to 5 dB below the music.) The VU meter was never intended to indicate peak distortions, or to indicate noise. Its advantages are in its ability to monitor mixed programme material, giving an apparent indication of loudness (which peak meters will not). VU meters are now being produced with LED indicators to indicate when peaks are reached.

The main attraction of the VU meter is its price. It is inexpensive, being merely a calibrated volt meter. However, many units do not reach the standard recommendations required and are often too fast and too 'bouncy'.

A VU meter is shown in Figure 14.6.

The peak programme meter

The peak programme meter (PPM) is not quite as old as the VU meter but its 50-year-old standards are still in use. It is much favoured in Europe, where it comes in varying national standards. It provides a more accurate assessment of overload and transmitter saturation, indicating programme peaks instead of

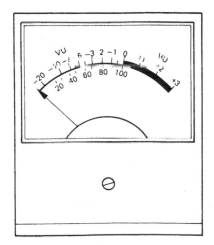

Figure 14.6 A VU meter (volume unit meter)

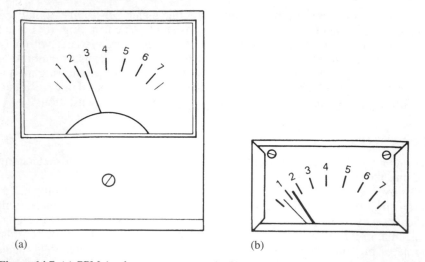

(a)   (b)

Figure 14.7 (a) PPM (peak programme meter); (b) stereo PPM (with two separate coloured needles)

their RMS value. It overcomes the shortcomings of the VU meter by holding peak levels, but it gives no indication of loudness. One form is calibrated between 0 and 7, with a 4 dB difference between each of the equally spaced gradations but 6 dB between gradations 0 and 1 and between 1 and 2. They all exhibit similar characteristics but with different calibrations.

When peak signals occur they are held in the circuit, and other information is ignored. In order to read the meter it is therefore necessary to wait until the peak hold circuit decays. The peak programme meter is normally lined up to 8 dB below the peak modulation point.

A peak programme meter is shown in Figure 14.7.

Other types of displays

Traditionally, needle meters are used to monitor sound levels visually, but other forms are available, such as bar graph displays, which use light segments that can be illuminated to show level. Workstations usually display peak bar graphs in reduced space, although needle-style meters are sometimes re-created. A workstation display of a peak reading meter is shown in Figure 14.8.

The colour of the display may change with high levels of modulation. Generally speaking, changes in levels and breaks in sound are easier to detect on this type of meter than on a needle meter. In addition, it is easier to see very low levels of sound on bar meters as they often show significant indications at 40 dB below peak. In contrast, a PPM cannot show much below 25 dB and a VU much below 12 dB.

Meter displays can be placed within video monitors but there are a number of drawbacks:

● Visibility is very dependent on the content of the picture.

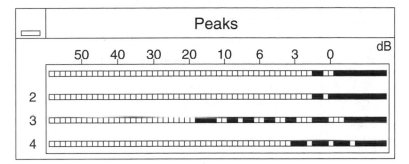

Figure 14.8 Workstation display of a peak reading meter. On overload the display turns red. The H key on the keyboard displays the meter horizontally, the V key vertically

- To reduce the complexity of the meters, some manufacturers do not include scale markings, and this can make these types of meter difficult to line up.
- Because of the need for burnt-in time code in audio post-production, meters have to be placed at the sides of the screen. With time code information and meters, TV monitors can become very cluttered.

Mono and stereo metering

Stereo sound recording systems require meters for each channel, with often an additional mono meter to monitor sound for the many monophonic systems that exist. It is normal to make visual meter and aural monitor checks by adding both signals together. In this situation, it might seem reasonable to expect the left and right signals to add up to roughly twice the value of the left and right signals together. However, in practice, this increase in level will vary with the degree of correlation between the two signals. If, for example, the two separate stereo signals were out of phase with each other, a combined mono signal would be less than either of the two individual signals. Only when the 'stereo' signals are identical, of equal amplitude, with 100 per cent correlation, such as a panned mono source in the centre of the sound field, equally divided between the two, will the combined signal be double that of the individual ones (an increase of 6 dB). Total correlation and no correlation are shown in Figure 14.9.

When there is no correlation between the channels, the sum level of equal amplitude signals is 3 dB higher than either input channel. Therefore in a worst-case situation there is a 3 dB difference between mono and stereo audio of equal amplitude. Different organizations take different attitudes to the problem. In Britain the BBC uses a 3 dB level drop (attenuation) for a combined mono signal, while NBC makes no attenuation at all. Much depends on the type of material being recorded (whether it is 'true' two-channel A/B or M & S stereo), the recording techniques and the correlation. In practice, the difference in sum level between mono and stereo falls somewhere between 0 and 3 dB.

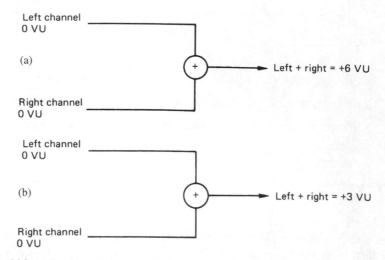

Figure 14.9 (a) Total correlation and (b) no correlation between left and right

Phase metering

In right and left channel systems, 1 dB of level difference between the left and right channels will be hardly detectable although a 3 or 4 dB error will cause the stereo image to be slightly one-sided. This condition does not cause a major problem for stereo listeners, but should the discrepancy be caused by phase errors between the two signals, this can produce difficulties for mono listeners. Errors in phase are essentially errors in time between two sources, caused, for example, when an analogue stereo replay head is misaligned at an angle to a standard tape. In this situation, on replay audio on one track is heard fractionally before audio on the other. The different timing is expressed as an angle. The more complicated and longer the programme chain, the more chance there is of phase problems building up in stereo. Remaining in the digital domain reduces phase errors to a minimum.

Table 14.1 lists the loss in dB at sample frequencies for a given phase error at 10 kHz.

In analogue stereo recording, phase problems first build up with incorrect alignment of the heads. Film and multitrack systems can suffer particularly badly from head misalignments, since phase problems are critically affected

Table 14.1 Chart showing loss (dB) at sample frequencies for a given phase error at 10 kHz

Frequency	15°	30°	45°
		[Loss (dB)]	
1000	0.00	0.00	−0.01
2000	0.00	0.00	−0.02
5000	−0.01	−0.06	−0.16
10000	−0.06	−0.29	−0.68
12500	−0.11	−0.46	−1.08
16000	−0.18	−0.28	−1.83

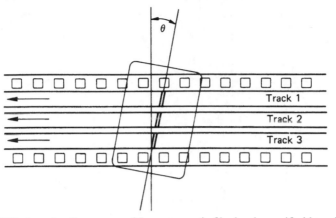

Figure 14.10 An azimuth error on a 35 mm magnetic film head magnified by using tracks 1 and 3 creating greater phase problems than by using adjacent tracks

by the distance between tracks. In digital systems the accuracy needed to process audio rarely creates phase problems on their own. However, problems can occur once the sound is removed out of the digital domain.

Figure 14.10 shows an azimuth error on a 35 mm magnetic film head.

To ensure that phase problems can be quickly discovered, reference phase tones on tapes should be copied from recording to recording, rather than be regenerated at each stage. Programme material with mono information can be usefully used to confirm phase condition. Centre-located dialogue from a mono source (having left and right signals of equal amplitude and in phase) is particularly useful in picking out phase errors.

Phase oscilloscope

The most straightforward and simplest way to display the phase relationships between two signals is to use a meter, but this is merely an indication of phase and only displays large phase problems. An oscilloscope display is a far more satisfactory device. The left signal produces a vertical deflection, while the right signal has a horizontal deflection. When the signals are perfectly in phase, and of equal level, a line slanting 45° to the right will appear. As the in-phase condition changes, ellipses will appear on the display.

Steady tones without phase error as seen on an oscilloscope are shown in Figure 14.11.

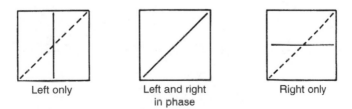

Left only   Left and right   Right only
                      in phase

Figure 14.11 Steady tones without phase error as seen on an oscilloscope

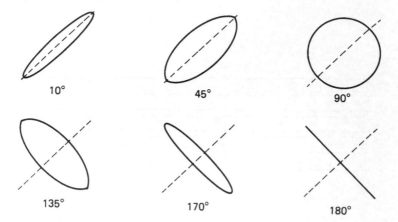

Figure 14.12 Steady tones with phase error

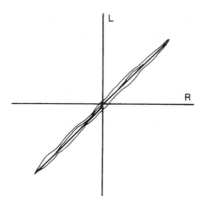

In phase

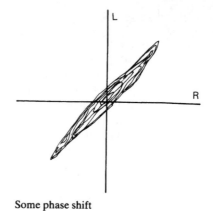

Some phase shift

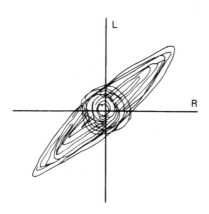

Increased phase shift

Figure 14.13 Phase shift

When the display shows a circle, there is a 90° phase difference between the two signals.

Figure 14.12 shows steady tones with phase error.

This type of meter not only shows the initial alignment of a stereo system when playing back alignment tones, but it also gives a continuous evaluation of phase during the recording. In the correct in-phase condition there is a narrow straight line of 45°. The sound image is in the centre. Switching from stereo to mono means that the image does not move.

Figure 14.13 shows increased phase shift.

When there is a moderate amount of phase shift, switching between mono and stereo produces a barely perceptible change in frequency content (a slightly enlarged line). This will be heard as a slight high-frequency roll-off.

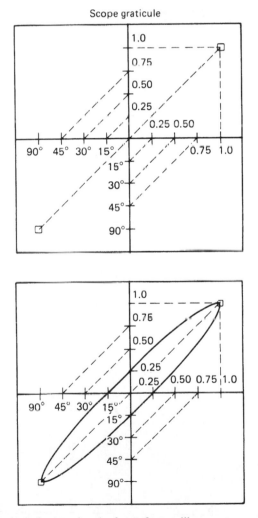

Figure 14.14 A graticule for covering the front of an oscilloscope to measure phase errors in stereo programmes, with an example of a 15° error between the channels

The high frequencies will be more out of phase than low frequencies if the fault is due to azimuth head misalignment.

Further shifts in phase cause a circular pattern in the centre of the display. This could be derived from voice sibilance. If the sibilance is out of phase between the two sources in the mid to high frequencies range, it will move to either side of the sound field, away from the main dialogue position.

An out-of-phase relationship of 30° is undetectable at 10 kHz, and since most material generally passes though a number of systems before release or transmission, a 15° out of channel phase relationship is often recommended as the maximum permissible at 10 kHz. This leaves an additional 15° margin for errors during transmission.

A graticule for covering the front of an oscilloscope to measure phase errors in stereo programmes is shown in Figure 14.14.

15

Mixing consoles and audio processing equipment

In the audio post-production studio the mixing console or desk or board is the main working area and the heart of the sound system. It is positioned to allow the mixer to hear the best sound, to view the picture comfortably, and to have good access to the various devices he or she needs to use during the editing and mixing.

In the large specialist studio where no soundtrack preparation takes place, only mixing (such as for theatrically released films), the mixer does not require access to the picture projector or to the audio machines, so these are usually placed in a sound proofed room close to, but away from, the mixing environment.

However, in most audio post-production environments not only does mixing take place within the studio area but also sound editing and track laying. In this environment all of the necessary equipment used must be kept within easy reach of the operator. Most equipment is quiet and this is not a problem; however, the video player may be noisy because of the cooling fans or spinning heads and this may need to be in a separate room. This is not a problem when mixing sound, for once the VCR has been loaded and put into remote operation, it is unnecessary to touch it (for EDL work and track laying it may be easier if the VCR is within the suite). The computer processing equipment can also be positioned away from the mixing area. This is an advantage if the computer controller serves other areas but cable lengths may need to be kept short if digital and data equipment is to function properly (lengths up to 30 metres are recommended for AES/EBU digital audio lines but only 10 metres for S/PDIF (Sony/Phillips Digital Interface Format) cables). Other machinery in the audio post-production suite, such as CD players, R-Dat players and sound processing gear not built into the mixing console, needs to be close at hand too. Such equipment is called outboard equipment, equipment within the console being known as inboard.

The mixing console

Essentially, the mixing console is a device for collecting together various sound sources, processing them, and eventually producing a composite output for recording. This output may be in mono, two-track stereo or multitrack stereo, to be used in motion picture, television or even video game productions. The ever-increasing demand for multi-channel formats has led to an increase in the number of channels needed and manipulated. Digital sound recording consoles with automation have made this much more viable for one-person operations and in addition, the audio quality is excellent.

In audio post-production the sound mixer's time is spent in viewing the pictures, and combining this with movement of the hand on the mixing console to produce the desired sound – it must be easy to translate 'thought from perception into action'.

This is most quickly achieved by physically moving faders and controls in the traditional manner. But many audio workstations include their own sound mixing screens within their systems where a mouse or keyboard controls the audio. This can be satisfactory in a music session where controls may be set and then left throughout a mix, but in audio post-production controls may need to be altered all the time and at the same time – by at least 10 fingers of one hand! Unfortunately a mouse takes up the whole hand, which is very limiting, allowing only one fader or parameter to be adjusted with the cursor at any one time.

A fader display for mouse control is shown in Figure 15.1.

This problem can be partly solved by using a sophisticated Midi 'controller mapping' device as an add-on. This can provide the opportunity for a stand-alone mixing console of faders laid out in the traditional way to be added to a workstation. Conventional mixing can now take place; fader moves are 'mapped' (followed) by the workstation's own graphical faders on screen.

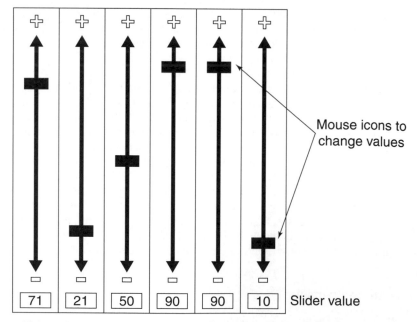

Figure 15.1 A video display of faders shown on the screen of a workstation – for mouse operation

These digital devices also remember and recall the positions of the faders as they are changed through the mix – mix automation. Being digital they maintain the integrity of the digital sound. In most situations standalone consoles are more practical and some workstation manufacturers offer these as specialist options.

Figure 15.2 shows an automated channel fader used on an integrated workstation.

Digital audio workstation sound consoles differ in their sophistication; even small standalone digital consoles often offer more facilities than their integrated workstation brothers. Manufacturers who make both workstations and recording consoles offer the most sophisticated integrated devices but at a high cost – perhaps three times the price of a high-quality broadcast digital recorder.

Programme material is mixed to picture as a project progresses; programme level is controlled by meters, perhaps the key to good reproduction. They need to be viewed at a glance and ideally should be in line with the picture screen or be part of the screen; similarly, all the controls should be close to hand and easy to find, without having to glance down, away from the screen, where action and sound are being related.

In designing traditional consoles, it is important that the manufacturer makes the controls used neither too small to be difficult to grasp, nor so large that they take up too much space. A compromise must be reached and this can

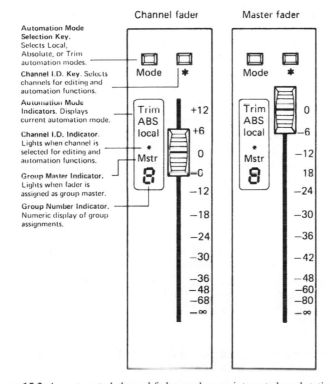

Figure 15.2 An automated channel fader used on an integrated workstation

be difficult. Small controls increase the time taken to make accurate settings (it has been demonstrated that a control knob of two inches in diameter is the ideal size for operation); but large controls give consoles so large a footprint (surface area) that it is necessary to have 'enormous extended' arms to use them! The length of consoles can be reasonably easily controlled; but it is the back-to-front distance that can be a particular problem. This is where the main controls of the console are to be found. Each sound source has its own fader to control the volume and further controls to adapt the incoming sound by changing the audio frequency response and by sending the sound to other sources. From the design engineer's point of view, consoles using digital sound processing can be more easily configured than analogue consoles. Modern digital signal processing has made these digital consoles cheaper and more versatile than their analogue counterparts. A 40-channel fully automated console can be purchased for little more than the cost of an industrial digital video recorder. In the most versatile digital desks it is possible for options to allow the end user to build his or her own console in software form almost from scratch. The custom-built console, which became too expensive a luxury in analogue form, is available again with all the advantages of digital sound.

Whatever the design, each channel will have its own fader, and within a console many of the sound channels, or input modules, will be similar. An optimum width for each strip is reckoned to be 40 mm, allowing the hands to cover five faders with comparative ease – most are smaller, because of space limitations.

The back-to-front distance of a console can be reduced (particularly in digital consoles) by allowing the controls to double up for various functions, which can be selected and shown in detail on a channel's own video display unit. Alternatively, one set of functions, shown perhaps in detail on a video display, can be made assignable to the various channels as required, although this means that only one function can be operated at once.

The number of input channels required for an audio post-production console depends on its purpose. In a small mono or stereo operation, perhaps using a modular eight-track recording system, only 12 channels may be necessary on the mixing console. In a specialist motion picture theatre where three mixers are sitting at a console, 70 or 80 automated channels may be required.

A basic, simple audio console of modular design with two groups into a mono output is shown in Figure 15.3.

Types of console

Two types of consoles are used in audio post-production: simple consoles where the audio from mono to surround sound is mixed and routed straight into the recording system, and those consoles which offer more sophisticated monitoring facilities such as those used in laying soundtracks.

Split consoles

Split consoles have their sound monitor controls positioned together as a particular function of the console. Usually these are to be found above the output faders.

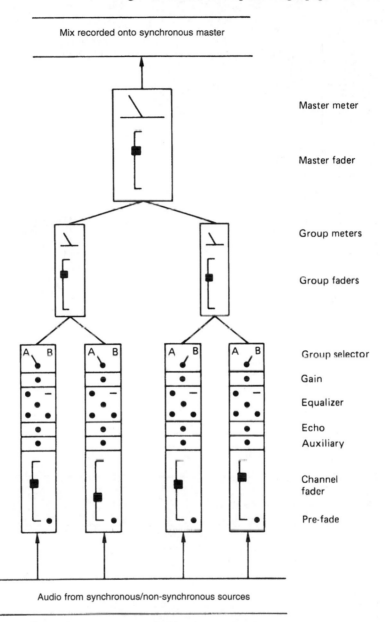

Figure 15.3 A basic, simple audio console of modular design with two groups into a mono output

In-line consoles

In-line consoles use what are known as in–out modules (Figure 15.4). These contain both channel controls and monitor controls. The audio signal is sent to a specific track on the multitrack recorder, and this track is returned on

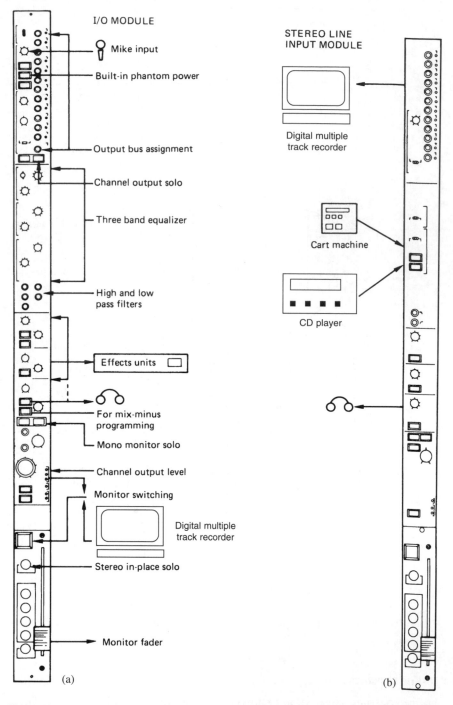

Figure 15.4 (a) An analogue in–out module for multitrack operation. (b) Stereo analogue input module with limited facilities designed for sources that are already equalized (*Courtesy of Sony Broadcast and Communications*)

playback to the same channel. Obviously, this takes up less space than the split console design but it does add more controls to the channel module. These types of consoles are regularly used in the more elaborate music recording set-ups and are found in audio post-production work.

However, they provide more facilities than are often needed, for unlike music recording, where a large proportion of the tracks may be recorded at once (or a large number of microphone inputs routed to a few tracks), in mixing sound tracks in post-production, sounds tend to be merely sent to one or more specific tracks at a time, in a similar fashion to audio workstations. Completely flexible channel-to-track switching allowing any module to be sent to any track is unnecessary, and can be replaced by group routing where sound is fed to a specific master control module. These groups, positioned at the centre of the desk, can then be dedicated to particular functions, such as music, effects and dialogue, and routed to the appropriate tracks.

A 20-channel, eight-group console is shown in Figure 15.5.

Inputs and controls

Sound is delivered to a mixing console at microphone level, at a standard line level, and in a digital audio console at various digital interfaces. In audio post-production consoles, there is only a limited requirement for microphone inputs. Most of the material comes from prerecorded sources such as tape or hard disc recorders, CD players or the multiple tracks from an audio work-station. If these pieces of digital equipment are to maintain their highest quality, it is essential that the digital path is followed. All should be synchro-nized to a master generator (in a small installation this is likely to be the recording console or workstation), sometimes called the wordclock master, which generates a sync pulse and allows the other devices connected to it to determine where the start of each digital 'word' begins. If any device is not

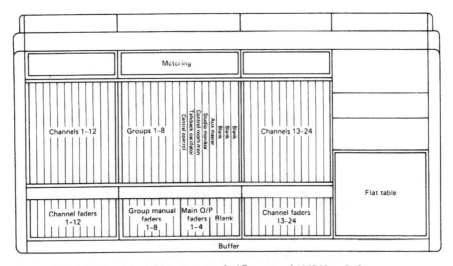

Figure 15.5 A 24-channel, eight-group console (*Courtesy of AMS Neve Ltd*)

synchronized to the system, dropout and other noises or glitches may well appear. Well-designed recording consoles will offer the necessary options for various digital sound interfaces, AES/EBU, ADAT etc. and sampling rates of 44.1 kHz, 48 kHz or more with quantization of 16, 20 or 24 bits.

In operations the signal is first passed through a volume control or fader, to adjust the level of sound. In mono, this is simply routed to the desk's output, but in stereo and surround sound, not only can the level be altered, but the sound can also be moved around the sound field. This movement is executed by means of a panoramic potentiometer, each channel having a 'pan' control which moves the sound image between the selected loudspeakers. In its simplest form the 'panpot' is a rotary control that moves the image across the two speakers. In a multitrack mixing system, a joystick (Figure 15.6) or mouse allows the sound to be moved around the four- or six-channel sound field. This facility may be selectable rather than available on every channel as panpots might be. These positions and other parameters may be capable of being automatically memorized within the mixing console itself – console automation.

Computerization of mixing consoles

The vast complexity of modern recording consoles has led manufacturers to provide computerized aids to mixing, making it possible to memorize every setting at a particular point in a mix as it progresses. Since a sound mix is created against time it is essential that the mixing console is tied to time. In audio post-production this will be the time code on the picture master, which could be either SMPTE/EBU or Midi time code through an interface. It is important that any automation system is capable of recalling all the necessary

Figure 15.6 Joystick controls for a multitrack surround sound system

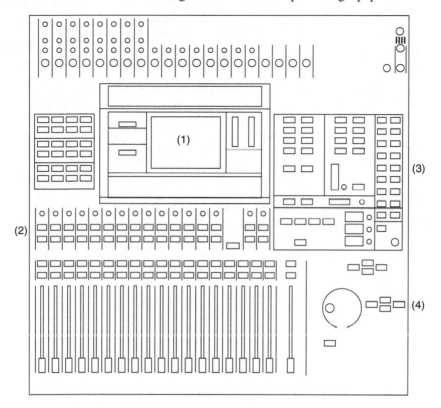

Figure 15.7 A small digital 20-fader fully automated console with central video display panel for routing equalization, scene memories, etc. (1) Video display unit; (2) equalization buttons; (3) control panel; (4) control for video display (*Courtesy of Yamaha Ltd*)

information within the time it takes for the workstation or audio machines and the video machine to reach synchronization. A system is of little use if the picture and soundtracks are up to speed and running in lock, but the automation system is still trying to synchronize. Two megabytes may be needed to memorize the automation on a 40-input recording console. Automation should never hamper the creativity of the mixer. In an ideal system, the operator would need to use no extra controls. Unfortunately this is not completely possible, but it is possible to go a long way towards this goal, much depending on the type of memory storage used and whether it is a reset or a recall system.

Recall merely tells the operator where the controls should be after a successful mix, but it does not reset them to the positions. Recall information is usually in the form of a video display screen, indicating, for example, where equalizers should be placed to correctly reproduce their previous positions. This facility is found in less expensive analogue consoles but not in digital consoles which can offer full reset automation. Digital signal processing can enable all the facilities on a console to be automatically reset, perhaps remembering 6000 different settings.

A small digital 20-fader fully automated reset console is shown in Figure 15.7.

This total automation system means that sound mixes can be stored in a console's memory system complete, without ever having actually been committed to tape or disc. These are called virtual mixes – the recording of the automation, consisting of all the console moves. Together with the actual soundtracks, this data can then be kept for a later date to be replayed and transferred as a final mix. Reset automation remembers faders' physical positions and puts the motorized controls back to their memorized points. The smoothness of this movement will depend on the processing power of the console. In less expensive consoles, changes of sound levels may be reset as a series of audible intermittent increments – which is exactly what they are! Much will depend on processing power and cost.

The data that records a computerized mix may be either memorized on a floppy disc, or recorded as digital information within the console's own long-term memory. These console 'instructions' can then be stored with the master audio and its components. The precise accuracy of automation is not vitally important in mono, since each individual sound source is controlled independently. However, in stereo the situation is different, for it is vital that a stereo pair of controls move together or else the stereo image will move across and around the sound field as mis-tracking takes place.

Digital consoles are able to offer complete automation of all their facilities. The ease of routing signals within them means that any function or parameter can be routed anywhere and be remembered.

Some automated display screens are shown in Figure 15.8.

Traditionally, consoles have additional facilities placed above each fader, physically available on the channel unit or strip . Digital consoles can use an alternative where an enter button is offered for each facility, giving each channel strip access to one set of assignable master controls. These may be displayed on a video screen, giving details and snapshots of mix settings and operating status. Alternatively, each channel may have its own video graphic display which can be selected to show any of the functions available on the channel. To make such systems easy to use, it is essential that there are not too many separate 'layers of functions' to go through to find the operating page for a particular purpose.

The other facilities needed for audio post-production on the input channels, either through assignment or through being physically available, include the following.

Auxiliaries

Each input module is likely to have various separate outputs (after the fader but prior to equalization) in addition to its normal outputs. These are called auxiliary outputs and are used to feed additional auxiliary equipment such as reverberation and delay units, as well as sending sound to other sources solely for monitoring purposes. In post-synchronization (the re-recording of poor-quality location dialogue in a studio), it is necessary for the artiste to hear the original dialogue as a guide track; this can be sent via the auxiliaries (sometimes called foldback) to the artiste on headphones. Auxiliaries can also be used in less sophisticated consoles to monitor tracks already recorded while laying down other sounds during the track-laying process.

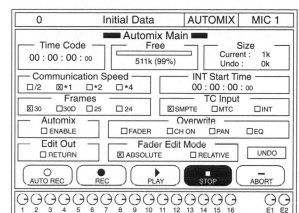

(a)

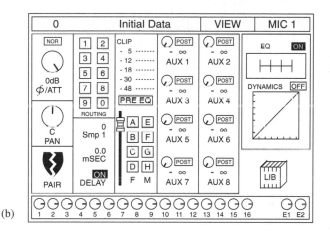

(b)

Figure 15.8 An automated digital console: (a) automation display; (b) snapshot of a scene display (*Courtesy of Yamaha Ltd*)

Auxiliary returns

Some signals that are sent via the auxiliaries need to be returned to the console after, for example, the addition of echo. These processed sounds can be connected to an input channel or, alternatively, returned via a special auxiliary return, in the form of a stereo input, together with a level control, a pan control, and routing selection.

Audition

This is another form of auxiliary but one specifically for monitoring. It includes much of the audio chain; for example, reverberation and equalization.

M/S switches

These allow M & S matrix stereo format recordings to be processed within the desk itself. These are usually dialogue tracks.

Mutes

Mute switches are cut buttons that silence a channel and are often used to cut out background noise when audio signals are not present.

Phase reversal switches

These merely reverse the phase of the input connector. They can provide quick confirmation that stereo is present.

Prefade listen (PFL)

Pressing the PFL button on a channel allows the channel to be heard before the fader is brought up. This can be used to advantage if tracks have unwanted sounds on them. These noises can be heard on PFL with the fader down and lifted when the problem has passed. (AFL is after fader listening).

Solo

This is similar to PFL but when operated, this mutes all other channels, making it particularly easy to hear the selected audio channel.

Width controls

These give an apparent impression of greater width to the stereo image by feeding a certain amount of the material, out of phase, back into the channel.

Equalizers

Simple equalization consists of treble and bass controls for changing the frequency response of the sounds. Equalizers are invariably used in audio post-production to match sounds. They can also be used to give sound apparent distance or perspective; for example, as a performer walks away from a camera his voice will reduce in bass content. This effect can be re-created by careful use or equalization.

The equalization or 'eq.' can be made more sophisticated by offering more frequencies for adjustment, and by adding control of the width of the frequency band used. This is known as the Q. The Q frequency can be switchable to various selectable values, or it can be continuously adjustable with the variable centre frequency. In the latter form the equalizer is called a parametric equalizer.

In audio post-production, switched rotary knobs are ideal since they can be quickly and accurately repositioned during a mix to match a previous setting. The more complicated the equalizer, the more difficult it becomes to reproduce position successfully. Audio post-production consoles tend to use

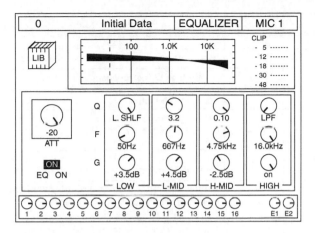

Figure 15.9 An automated digital console: equalization parameters. The equalization function is selected at each channel (*Courtesy of Yamaha Ltd*)

simpler forms of equalization because of this, although with computer memories to aid mixing more complicated equalization is now possible.

Figure 15.9 shows the equalization parameters of an automated digital console.

There was once a general philosophy in the broadcasting and music industries that frequency response correction should only be used to correct the deficiencies of equipment, rather than to modify sound, so the term 'equalizer' was applied. This philosophy has changed, but even so equalization should only be used with caution. The ear only too easily becomes accustomed to a particular unnatural 'quality' of sound which it then considers to be the norm.

Figure 15.10 shows an equalization display on a workstation screen.

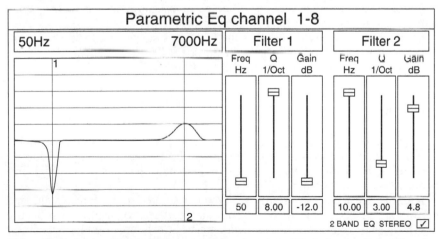

Figure 15.10 An equalization display on a workstation screen. On the keyboard, the C key displays the curve and the E key displays the eq parameters. Figures can also be typed in. The mouse can also be used to adjust the equalization curve

Filter type	Symbol	Application
Parametric	⟨⟩	Boost or cut can be set for a variable width frequency band centered around a selected frequency.
Notch	∨	A very high Q (narrow band) filter with a very large (effectively infinite) cut value. The gain control is not used with this filter.
Hi shelf	⟨	Allows gain or cut to be set for all the region *above* the boundary frequency. The Q control is not used with this filter. 12 dB/octave.
Lo shelf	⟩	The Lo shelf is like the Hi shelf, except that the controllable region is *below* the boundary frequency.
Hi pass (Lo cut)	⌐	A filter with a 12 dB/octave rolloff *below* the selected frequency. The Gain and Q controls are not used with this filter.
Lo pass (High cut)	⟍	Like the Hi pass, except the rolloff is above the selected frequency.

Figure 15.11 Graphic representations of equalization curves as used in some video displays on automated consoles

There are other forms of equalization found on sound consoles.

Notch filters

Notch filters used in mixing consoles have a high Q and allow a specific frequency to be selected and attenuated. They are used to reduce such problems as camera noise and can be 'tuned' to the appropriate frequency. The offending noise can often be removed without completely destroying the basic sound quality. A depth of at least 20 decibels is necessary for successful noise rejection. Tuned to 50–60 Hz rejection, the filter can reduce mains frequency hum, with a further 100–120 Hz notch for its audible second harmonics. A 15 K notch filter can be used to reduce the 625 line sync pulse interference noise radiating from domestic televisions in Europe.

Figure 15.12 shows a selection of notch filters.

Pass filters

Low-pass filters restrict the high-frequency response of a channel, allowing low frequencies to pass through. They are useful for reducing high-frequency electrical noise from fluorescent lamps and motor systems as well as high-frequency hiss from recording systems.

High-pass filters restrict the low-frequency spectrum of a system and permit high frequencies to pass. They are particularly useful in reducing low-frequency rumbles such as wind, traffic noise, and lighting generator noises.

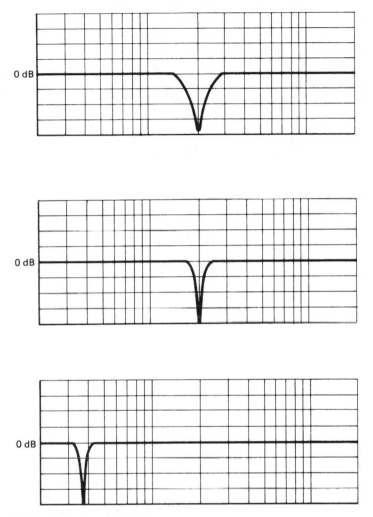

Figure 15.12 A selection of notch filters

Graphic equalizers

Graphic equalizers are able to provide equalizing facilities over the entire frequency range, and are often purchased separately from the console as outboard equipment.

They use sliding linear faders arranged side by side. This gives a graphical
representation of the frequencies selected; hence the name, graphic equalizer.
The individual filters are usually fixed and overlap.

They give quick and precise visual indications of the equalization used. As
with most frequency response filters, graphic equalizers should be provided
with a switch to remove the filter in and out of circuit, allowing the operator
to check whether the effect being introduced is satisfactory.

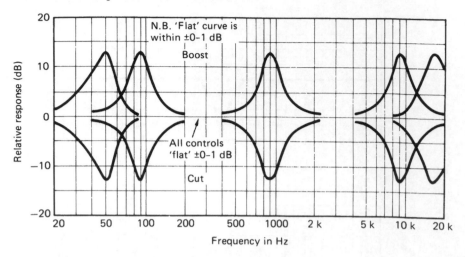

Figure 15.13 A graphic equalizer with selectable frequencies at 50, 90, 160, 300, 500, 900, 1.6 k, 3 k, 5 k, 9 k and 16 khz (*Courtesy of Klark-Technik Research Ltd*)

Uses of equalizers

Equalization can be used for various reasons:

- To match sounds in discontinuous shooting where the distance of the microphone from the performers varies as the camera shot changes; with equalization, it is possible to match the dialogue, producing a continuous cohesive sound quality.
- To add clarity to a voice by increasing the mid-range frequencies.
- To correct the deficiencies in a soundtrack if, for example, a voice has been recorded under clothing and is muffled.
- To add aural perspective to a sound that has been recorded without perspective.
- To remove additional unwanted sounds such as wind noise, interference on radio microphones, etc.
- To produce a special effect such as simulating a telephone conversation or a PA system.
- To improve the quality of sound from poor-quality sources such as transmission lines or archive material.
- To improve sound effects by increasing their characteristic frequency, for example by adding low frequencies to gunshots or punches.
- To 'soften a hard cut' (upcut) by reducing a specific frequency at the start of a clip and then returning it to normal; this is less harsh than a fade.

Control of dynamics

The range between high and low levels of sound, in reality, is often far greater than is needed for a sound mix. The simplest way to control this is to pull the fader down during loud passages (where overload of the transmitter or the recorder could take place), and to push it up during quiet passages (ensuring that sounds are not lost in the background noise of the system or more likely the reproducing environment) (Figure 15.14).

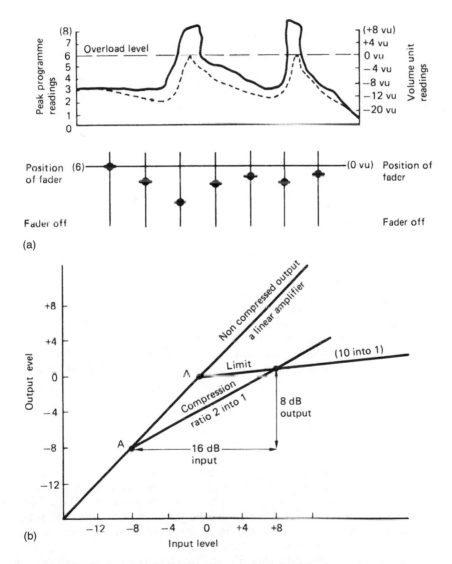

Figure 15.14 (a) Adjustments to a sliding linear fader in manual compression of audio; (b) sound-level diagram showing the effects of limiting and compression, 'A' being the breakaway threshold point from the linear

This manual dynamic control can work effectively, but it is often much simpler and easier to control dynamics automatically, using a compressor/limiter. More and more of these are being built into the channel modules of consoles, allowing each channel separate limiting and compression, and often with automation of the parameters.

However, more normally they are found, and with greater sophistication, outside the console as 'outboard equipment'. They are available from various manufacturers, some of whom have even returned to using the tube or valve with its ability to produce a less 'harsh' sound in compressed and limited material.

Limiting

Limiting is used to 'limit' or reduce a signal to a specific level. Further increases in input level result in no further increases in the output. The severity of the reduction is measured as a ratio, usually of 10 to 1, although some limiters can give ratios of 20, 30 or even 100 to 1 (the difference between these ratios is audibly not great). Some limiters may allow fast transients (that is, short fast rising signals) to pass through without being affected by limiting action. This maintains the characteristic of the sound without the limiting effect being audible. Limiting is an invaluable aid in digital recording where an overload could lead to a distorted and unusable recording.

Compression

Compression is a less severe form of limiting that is used to produce certain specific effects as well as to control and limit level. The onset of the limiting effect is smooth and progressive. The threshold point is the point at which compression starts, ratios being between 1.5 to 1 and 10.1.

Low compression ratios operated at low threshold points will preserve the apparent dynamic range of programme material (despite compression). However, at the same time, they will allow a high recording level and thus give a better signal to noise ratio. A high ratio at a high threshold point gives similar results but with the probability of a more noticeable limiting action.

The speed of the attack of incoming signals into a compressor or limiter is called the attack time. The quality of the compressed sound is very dependent on the speed at which the compressor attacks the incoming sounds. Slow attack times will result in a softening or easing of the sound. As the attack time lengthens, more high frequencies will pass unattenuated through the system; on speech this will lead to sibilance. Slow attack times are useful when a considerable amount of compression is needed and, when used with a tight ratio, low-frequency sounds will have maximum impact. With deliberate overshoot these sounds will have added punch, which can be useful in recording sound effects. Faster attack times are necessary for speech and can be used to assist in controlling apparent loudness.

With very fast release and attack times and high ratios of compression, the low signal content of programme material is raised. This produces a subjective increase in loudness and is particularly useful, if carefully used, in increasing dialogue intelligibility. Unfortunately, if an extremely fast release time is used, a pumping or breathing effect becomes apparent as the background level goes up and down in volume with the sound.

A traditional analogue compressor limiter is shown in Figure 15.15.

Recovery times can often be set automatically in compressors, being dependent on the level of the input signal. A specific recovery time is automatically programmed when the signal reaches above a certain threshold. As soon as the input falls below this threshold level, the recovery time smoothly changes to a shorter one, perhaps from 10 seconds to one or two seconds. This is sometimes referred to as the 'gain riding platform'. It is used in some broadcasting transmitters where considerable overall long-term compression is needed, and can affect the sound of carefully mixed programme material.

When compression is used on an entire mix, it is possible to end up with one dominant signal on the track. This can be a particular problem when using loud sound effects, or music under dialogue, and may even show itself as pumping or breathing. It is therefore better to compress the various sections of the mix separately, rather than all at once. Effects and music under dialogue can even be held down with a compressor rather than with a fader.

Certain compressors are capable of splitting compression into various frequency bands, and this can be particularly useful with dialogue where the power varies at different frequencies and unnatural effects can be reproduced. These occur at the lower end of the frequency components of speech that form the body of words, and give character.

Normally compressors incorporate gain level controls, allowing levels to be maintained even when gain reduction takes place. In this way a direct comparison can be made between the compressed and the uncompressed material. In stereo, compression can create movements in the stereo image and they should be used carefully.

Noise gates and expanders

Noise gates and expanders are the inverse of limiters and compressors. A compressor is used to reduce dynamic range, whereas a noise gate is used to increase dynamic range by reducing the quieter passages further. The point where the reduction in level occurs is called the 'gating threshold', and this is adjusted to just above the unwanted sounds.

Figure 15.16 shows a frequency selective noise reduction device operating at four frequency bands.

Normally inserted at the input to the channel, noise gates must be used with care, since low-level sounds that are required for a mix, such as whispered dialogue, can be treated as unwanted noise and also reduced in level.

Some noise gates are available as frequency selective devices, so that each band of the frequency spectrum can be individually noise-gated. This is ideal for reducing camera noise and other unwanted sounds recorded on location. The attack times and release times of the gate must be as short as possible, to minimize clipping on programme material. Expansion can also be used as a method of noise suppression, by exaggerating the difference between the wanted and unwanted sounds.

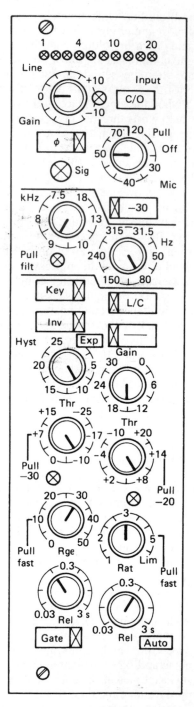

Figure 15.15 A traditional analogue compressor limiter with gate/expander facilities. Hyst: hysteresis gives a greater control of compression; Inv: inverts the noise gate to a ducker (*Courtesy of AMS Neve Ltd*)

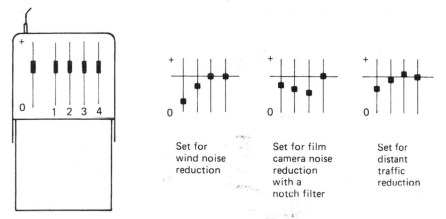

Figure 15.16 A frequency selective noise reduction device operating at four frequency bands

Limiters and compressors are used to:

- provide overload protection against over-saturating; this is particularly important in digital or optical recording and to a lesser extent in analogue recording using metal recording tapes, where over-saturation leads to instant unacceptable clipping;
- reduce the dynamic range of material to make it more suitable for the final format (non hi-fi video cassettes have a very limited dynamic range);
- automatically reduce the range of sound to a comfortable level, for domestic consumption;
- increase the apparent loudness or create impact;
- increase the intelligibility of speech and the 'definition' of sound effects;
- reduce background noise levels, in noise gate and expander form.

Reverberation

Other facilities provided on the console include units designed to add artificial reverberation. Reverberation occurs when sound waves are reflected off surfaces – this is not the same as echo, which is a single reflection of a sound. Reverberation adds colour, character and interest to sound and sometimes even intelligibility. Reverberation accompanies nearly all the sounds that we hear, and its re-creation can be a very vital part of a sound mix, particularly in speech when quite often sound that is totally lacking in reverberation seems unnatural. Natural reverberation helps to tell the ears the direction that sound is coming from, the type of environment that the sound was recorded in, and approximately how far away it is. It is particularly relevant in stereo recording, but within multi-channel formats artificial reverberation must be used with care. Mixes with artificial reverberation can suffer from phasing problems when mixed down to a mono format.

Clapping the hands gives an excellent indication as to the reverberation of an environment, and is often used by sound recordists when they first encounter a new location.

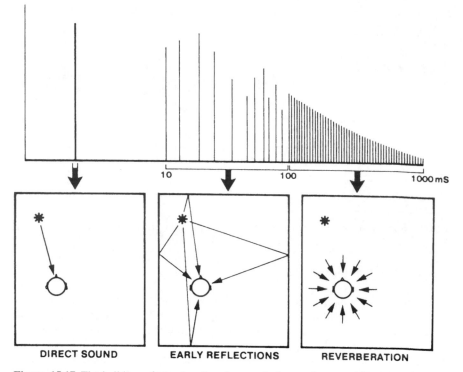

Figure 15.17 The build-up of reverberation from a single sound source (*Courtesy of Klark-Technik Research Ltd*)

When the hands are clapped, sound radiates in all directions at a rate of about a foot per millisecond. The first sounds to reach the ears come directly from the hands, and tell the brain where the sound source is. The next sounds heard come from early reflections, and these will be slightly different from the original sound waves since some of this sound energy will have been absorbed by the surfaces that the sounds have struck (this absorption will depend on frequency). These early reflections can extend from 5 ms, up to about 200 ms in a large hall. These reflections will build up quickly to an extremely dense sound, coming from all directions – no longer just from the hands. The ears now receive a slightly different pattern of reflections, at different times, from different directions. The time this effect takes to die away is called the reverberation period.

The build-up of reverberation from a single sound source is shown in Figure 15.17.

If an electronic device is to re-create reverberation satisfactorily it must take into account all these parameters.

Essentially, reverberation units take sound and delay it – often through digital delay lines. These delays can range from a few tens of milliseconds to a few seconds. The re-creation of the acoustic of a small room requires only a few milliseconds of delay, whereas a few seconds of delay is required to re-create a public address system. By taking different parts of a digital delay line and applying degrees of feedback and filtering, more sophisticated effects can be created. It is possible to de-tune the delay and, with time slippage, to produce an effect that will split mono sound into pseudo stereo. Perhaps more useful in audio post-production are pitch-changing and time-compression devices, where specific sounds can be made to fill predetermined time slots. An example might be where an additional piece of dialogue has to be added to an edited picture, or where a voice-over for a commercial is too long and needs to be shortened. Using a delay device, this sound can be fed into a memory and then read out at a different speed, either faster or slower, but at the same pitch. Reading out data slower than it is fed in means that the system has to repeat itself and there may be an audible hiccup or 'glitch' in sound.

Requirements for digital recording consoles

Digital recording consoles for audio post-production ideally need to have:

- all parameters automated;
- virtual mixing;
- compatible digital audio interfaces;
- comprehensive equalization and compression facilities with automation;
- integrated transport control and 'punch in' record facility panel;
- multi-channel sound capabilities;
- sufficient monitoring for the various 'stems' required;
- the ability to store 'snapshots';
- data removal compatible with the local systems used.

16

The mix

The audio post-production of a project is completed with the mixing of the soundtracks. The audio will have already been prepared, probably in the form of soundtracks on a digital workstation; some of the tracks may have already been mixed and their levels set; much will depend on the time allocated, and the experience of the sound editor, although in many situations the sound mixer will have also laid and prepared the soundtracks. In productions for theatrical release and television drama it is not unusual to employ a specialist sound editor or editors. However, the distinction between picture and sound editor and sound editor and mixer is blurring as equipment becomes more versatile. It has been argued that with the greater sophistication and efficiency of audio workstations, fewer tracks are needed in productions.

There are both aesthetic and technical aspects to sound mixing. Technically, the most important part of audio post-production is to produce a match of the sound to the visuals so that sound appears to come from the pictures. The visual and aural perspective must be right – if an actor moves towards the camera, tradition and realism dictate that the acoustic of his voice should sound as though he is moving towards the camera. If an explosion occurs half a mile away, it should sound as though it occurs half a mile away, regardless of how close it was originally recorded. Of course, all rules are there to be broken.

Aesthetically, audio post-production is concerned with mixing the various soundtracks to produce a cohesive, pleasing, dramatic whole, that enhances the pictures to the director's wishes. A scene, from an audio perspective, is more than just one sound following another (as are the visual pictures); it consists of sounds that knit whole scenes together, adding atmosphere and drama. Sound places pictures in geographical locations and indicates time. The sound of an aircraft passing overhead will join together different visuals to one geographical location – perhaps an airport. Unfortunately, sound is often considered very much a Cinderella compared to vision, and directors sometimes take little trouble to supervise mixing. Others will take sound into account from the very start of a production by using audio creatively to add drama.

Operation of the controller

During the mix, the soundtracks and the picture are controlled from the work-station, the picture recorder controller or a synchronizing controller. Originally in the 1930s, audio post-production systems were only capable of running in synchronization from standstill up to speed and into play. Once a mix had started, the 'take' had to continue for the length of the roll to a maximum of 12 minutes. If a mistake was made, it was necessary to unlace, rewind the soundtracks and return to the beginning for a new mix, or alternatively, to re-record the faulty sequence and cut it into the final mix. A successful mix was called a print (since it could only be reproduced via a photographic print from the optical negative), and this term is still used today in the film industry. With the advent of magnetic recording, it became possible to replay the sound immediately rather than have to wait for a print to be photographically made. Originally these systems only ran at single speed, but they were later developed to run in interlock backwards and forwards at single and fast speeds.

In audio post-production this operation is traditionally known as 'rock and roll'. The soundtracks are run forward and mixed; a mistake is made; the system is rolled back; it is then run forward again in synchronization; recording begins again at an appropriate point before the mistake. To make sure that the insertion into 'record' before the mistake matches the previous mix, a balancing key or send/return switch is used. This compares the sound that is about to be recorded (the send), with the previously mixed sound (the return). If, with the tracks running, the sound that is about to be balanced matches (balances) sound that is already on the track, when the record button is pressed or 'punched in' there will be complete continuity of sound on the final master. This facility needs to be specifically designed as part of a monitoring system itself. This facility is usually included in fully automated digital consoles where the mix can be 'recorded' as a memory of all the console's switching and fader movements (a virtual mix) rather than as an audio recording. This allows unlimited changes to be made to a sound mix with the final copy always just one generation away from the laid soundtracks.

Mixing using 'rock and roll' techniques can limit the continuity of a mix. The system is stopped and started every time a mistake is made; certainly rehearsals can reduce these problems, but all this takes up time. To help provide cues to mix a soundtrack, originally lines were physically drawn across a film at an angle. When the film passed through the projector gate the line was seen to finish at the audio cue. Today, workstation edit screens provide cues of all the soundtracks as they pass the time line and are heard at the monitors.

Console automation

With the introduction of console automation it has become possible for mixes to be memorized in their entirety. Originally designed for the music industry, console automation was confined to the automation of the faders. The automation of equalization and other dynamics was not essential. In audio post-production, fader-only automation is not really sufficient. Mixing is a somewhat different process from music recording; the quality of the sound-tracks is often altered to match the pictures. This will mean not only changes in level on the channel faders but also changes in frequency equalization. This can happen at almost every different take when dialogue is shot discontinu-ously – here each take in a dialogue sequence may have a different quality (due to varying microphone distance from the speaker). This means total automation of a console is desirable; this also allows 'snapshots' of parame-ters to be taken too. For example, an equalization parameter at a particular location can be stored and repeated accurately at other points in the mix whenever that location turns up. Routing and patching can also be recalled to repeat the set-up of a particular section or mix.

The virtual mix

In the total reset automated mix, the movement of the faders is memorized in exact synchronization with the soundtrack using the time code or the Midi interface of the soundtracks. Here it is unnecessary to actually 'record' the mix as it progresses – all that is necessary is to be able to store and repeat all the actual movements that the mixer makes on the mixing console. These set-tings are recorded as a data stream of the fader movements, equalization and other sound processing moves in synchronization with the picture.

Replaying the mix automatically recalls the console's memory and oper-ates all the controls in synchronization with the memory. The original audio is never altered; the operator is free to make updates. He or she merely over-rides the movements of the faders where necessary and makes a series of improvements to the mix until the final desired result is reached. No longer is it necessary to stop a mix when a mistake is made, and lose continuity. The mixer now merely returns to the faulty part of the mix and just updates the memory.

A screen display of fader automation is shown in Figure 16.1.

However, the mix can only remember audio parameters that are inter-

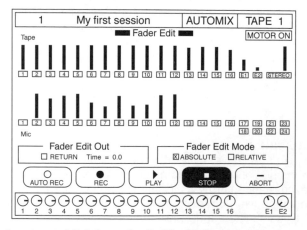

Figure 16.1 An automated digital console; display of fader automation

faced via the console's memory; if audio is fed out of the console and patched through audio outboard equipment, such as compressors, these will not be part of the console's mix. The movements on the console will be reproduced but no processing will be introduced unless the outboard equipment is again tied into the console's operation! Similarly, audio introduced from outside the system will not be reproduced at a later date unless it is deliberately introduced and synchronized to the system, perhaps by being recorded into the workstation. When a virtual mix is finally completed, with no possibility of further changes, it will be stored onto tape or disc.

Sound sources for the final mix

In a large audio post-production mix, such as for a theatrical film, tens of individual tracks may be used to make up the final soundtrack. In a small video production there may only be two tracks available, recorded on an industrial video recorder. Soundtracks for the final mix will have been provided from various sources:

- Original dialogue that was recorded in the field, which may be in mono or stereo, and will also contain some incidental effects as well as the spoken lines.
- Wild track dialogue that was also recorded in the field, not at the time of the shooting but immediately afterwards, and which has been fitted to the picture.
- Post-synchronous dialogue which consists of dialogue which had been recorded to picture after the shoot, recorded in mono and difficult to match to dialogue actually recorded in stereo on location.
- Post-synchronous effects which were recorded in the studio and not in the field (Foley effects), mono.
- Effects which come from an effects library or location recordings, mono, stereo and in specialist situations in surround, for sports events for example.
- Scored music especially composed and recorded to go with the production, recorded in all formats from mono to multi-channel.
- Source music which is intended to come from musical sources on or off the screen, recorded in all mono to multi-channel formats.
- Library music that has been taken from a music library and matched to the pictures, mono to multi-channel formats.

Cue sheets

To assist the mixer take accurate cues during the mix, cue sheets are sometimes provided by the sound editor or his or her assistant, showing the various available sounds in graphical form. A traditional cue chart is shown in Figure 16.2.

Once considered a necessity, these are now only used on the very largest productions. Modern workstation systems provide all of the information

Audio mixing chart

Time in seconds

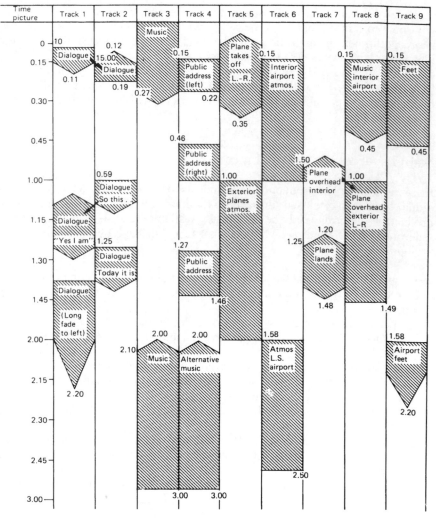

Figure 16.2 A traditional cue chart

needed to mix tracks. Most valuable is the time line across which the laid tracks pass. The tracks may themselves have information written onto them or show a graphical display of the audio waveform.

Figure 16.3 shows a workstation display screen showing parts of the project in Figure 16.2.

The sheet is divided into several columns (up and down or across) depending on the number of soundtracks that have been prepared. Each soundtrack has its own column with the first column giving visual cues and timing. The columns may be sub-grouped into dialogue, music and effects. As the mix progresses, the tracks are identified against the time readings on the chart. The

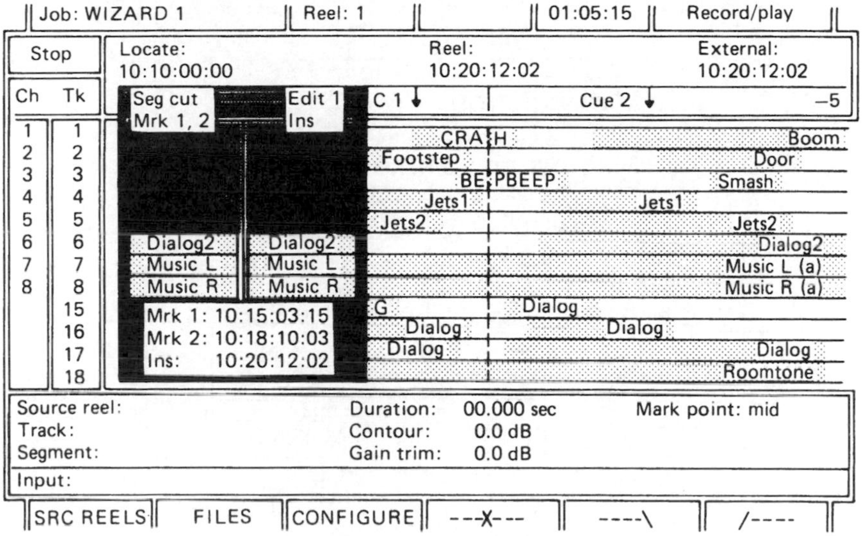

| Job: WIZARD 1 | | Reel: 1 | | | 01:05:15 | | Record/play | |

| Stop | Locate:
10:10:00:00 | | | Reel:
10:20:12:02 | | External:
10:20:12:02 |

Ch	Tk	Seg cut .... Edit 1	C 1 ↓		Cue 2 ↓	−5
1	1	Mrk 1, 2 .... Ins	CRASH			Boom
2	2		Footstep			Door
3	3			BEEPBEEP		Smash
4	4		Jets1		Jets1	
5	5		Jets2			Jets2
6	6	Dialog2 Dialog2				Dialog2
7	7	Music L Music L				Music L (a)
8	8	Music R Music R				Music R (a)
	15	Mrk 1: 10:15:03:15	G	Dialog		
	16	Mrk 2: 10:18:10:03	Dialog		Dialog	
	17	Ins: 10:20:12:02	Dialog			Dialog
	18					Roomtone

Source reel:		Duration:	00.000 sec	Mark point: mid
Track:		Contour:	0.0 dB	
Segment:		Gain trim:	0.0 dB	

| Input: |

| SRC REELS | FILES | CONFIGURE | −−X−− | −−−−\ | /−−−− |

Figure 16.3 Workstation display screen showing parts of the project in Figure 16.2

time counter within or below the picture provides film feet or time information tied to the picture, be it film or video.

The various soundtracks are mixed together with sounds cutting from one track to another, fading out and fading in or, alternatively, cross-mixing. Equalization, reverberations, compression and limiting are added as required. For complicated productions pre-mixes or pre-dubs are produced to help simplify the final recording – it would be impossible for one pair of hands, even with automation, to control 30 separate soundtracks at once! The route taken for mixing depends on the type of equipment used.

Audio workstations

Mixing soundtracks to produce a final soundtrack within the confines of a digital workstation (picture and sound or sound only) using only the screen and a mouse to manipulate audio is difficult. Using a mouse allows only one track parameter to be adjusted at a time. In some budget workstations there are other problems; adding equalization to a track within the system may only lock the equalization to that location and time – if the track is subsequently moved the equalization can stay where it is! With limited facilities, perhaps only a four-track system, it is simpler when laying tracks in the workstation to include as many as possible of the fades and equalization parameters that are known to be needed, leaving any uncertainties to the end. However, in the final mix the actual comparative levels of the dialogue, sound effects and music will be critical and it is sensible to divide and produce the tracks in three distinct groups, allowing reasonable control over the final mix. Even within a simple four-track system it is possible to produce a quality mono soundtrack, particularly if the tracks are carefully prepared. If the workstation

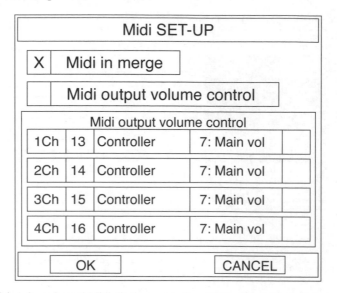

Figure 16.4 A four-channel Midi display screen for a Midi workstation interface. The output volume control is selected. If the controller messages are recorded into a Midi sequencer which is synchronized with the workstation, the entire mix and equalization parameters can be automated

is tied via Midi to a simple automated digital audio console, the system can work quite successfully and the mix will stay as a virtual mix within the workstation, ready to be updated as required or transferred to the video recorder. An alternative is to use a simple analogue mixer not tied to the workstation and bring the tracks out of the workstation and mix them directly onto a separate recorder or back into the workstation, but this time the result will be fixed and complete.

A four-channel Midi set-up screen display for a Midi interface with virtual mixing is shown in Figure 16.4.

In larger systems where a number of tracks are being mixed together, they may need to be sub-grouped into various smaller pre-mixes, for easier handling. Using a totally automated console or workstation, these can be held in the form of virtual pre-mixes of various tracks. In these systems the mix is memorized as a virtual mix locked by time code to the tracks. The actual tracks exist unaltered as different mixes are tried and recalled.

If a pre-mix is produced without using an automated console, it will be recorded onto a track or multiple tracks in lock with the complete system. It will then be replayed with other tracks for a further final mix to be made. (If the recorder is not a digital machine there will be an increase in noise.) Unfortunately the pre-mix is now set and it will now not be possible to alter it since the levels have already been decided and set. If the mix had been 'recorded' as a virtual mix, further changes could have been made. Much will depend on the sophistication of the mixing console as to the number of tracks that can be memorized as a virtual mix. In a theatrically released film many tracks and their various parameters need to be remembered, which requires an expensive console with a lot of processing power.

Pre-mixes can be divided down into groups such as dialogue, effects and music; or even further, into pre-mixes of spot effects, Foley effects, atmosphere effects, etc. Pre-mixes must be carefully handled, since it is possible for certain sounds to be obscured when the various pre-mixes are finally mixed together. Depending on the equipment, it may be possible to monitor other tracks while the pre-mix takes place. Stereo positioning will also be determined.

In many audio post-productions, the dialogue will be the most important part of the mix. A dialogue pre-mix is therefore often recorded first. This allows the mixer to produce the best possible dialogue quality, concentrating specifically on matching sound levels, frequency equalization and stereo positioning. Dialogue is rarely recorded on location in stereo (sometimes in Europe, rarely in the USA). The pre-mix will include panning information for stereo and possibly reverberation, but as many options as possible should be left open for the final mix if the mix cannot be altered later.

When pre-mixing it is often sensible to reduce the studio monitoring level occasionally to confirm that the pre-mixed sounds will not be lost when added to the final mix at a lower level. This can be a particular problem when heavy effects are being mixed under what will eventually be dialogue. Equal loudness curves show that we perceive sounds differently at different levels, so any pre-mix must be treated with care if it cannot be altered when the final mix takes place – only if the mix is held as a virtual pre-mix can it be easily altered.

Once the premix is completed the final mixing begins. Pre-mixes are, perhaps, not so necessary if an automated console is used.

Using an automation system, the mix for a documentary video programme on an airport might progress as follows.

Five or six various effects need to be mixed, plus dialogue tracks. In the first pass, the mixer feels that the background atmosphere mix is correct but the spot effects of the public address system are too loud, and that the plane passing overhead reaches its loudest point a little too late; perhaps the director doesn't agree with him. On the second pass, an attempt is made to improve the public address effect further, but this time the dialogue is slightly obscured. On the third pass, a slight attempt is made to increase the level and improve the intelligibility of the dialogue by reducing the background effects. This, the director likes. Now, all three attempts can be reviewed and compared by memory recall. It is decided that the final mix should consist of the atmosphere mix from the first pass and the dialogue mix from the second. Now the sound mixer merely selects the automation related to the appropriate fader and parameter movements from the computer, and the final mix is produced.

As the final mix is memorized, any last-minute updates of the balance of the two tracks can also be carried out.

A second approach is to use sections from the various mixes. The first 30 seconds from mix 1 is used, the next 3 minutes from mix 3, and the rest from mix 2, making the combined final track.

A third approach is to use the computer for virtual pre-mixing. Various soundtracks are taken and mixed as a pre-mix, and then further tracks are added, so building up a final track.

If a non-automated mixing console is used, the mix will progress more carefully; each scene or section will be mixed and agreed upon before the next portion of the soundtrack is mixed. If analogue recording equipment is involved, care will need to be taken to ensure that pre-mixes are recorded at as high a level as the recording medium will stand to ensure that the minimum amount of system noise is introduced. The pre-mixes are then mixed and copied onto the final master to produce final master mix. In a virtual mix, the final soundtrack may be held only as digital data of the varying console parameters and source audio. If this is held on a computer disc rather than magnetic tape, there is instant access to any part of the mix.

As the mix progresses it may become necessary to change the 'time position' of a soundtrack. Perhaps a piece of voice-over interferes with a sound effect, or perhaps the position of a sound effect is wrong in relation to the visuals. Using a workstation or computer disc modular multitrack recorder requires only to operate the slip mode of the appropriate track, making sure any processing, such as equalization, is locked to the sound move as well. Using other systems it may be necessary to transfer the sound to another machine and slip the second machine physically in relation to the synchronization of the first. Unfortunately, tape-driven modular digital multitrack machines only allow a few frames of slip in relation to the other digital tracks within their systems.

The final sound balance

The final sound balance of a production needs to conform to the local technical requirements, having:

- the correct relative sound levels;
- the correct dynamic range for reproduction;
- a consistent tonal quality;
- high intelligibility;
- the required perspective;
- the required acoustics;

thus making a blend of sound that achieves the desired artistic effect.

The final sound balance is the result of mixing together the various discrete elements that make up a soundtrack. Mixing together the final pre-mixes produces the final mix stems. The stems are the various strands that constitute the mix. In a large production these can be many tracks. The surround mix will be held as discrete, screen left, centre and right together with surround tracks rather than encoded into two tracks; other stems will be held as separate dialogue, effects and music tracks. Keeping the final mix as separate tracks makes it easy to prepare other versions of a production, which may eventually be a mono mix or a music and effects track.

Compatible mixes

By using level control, frequency equalization, varying dynamics and reverberation, the mixer has produced a final soundtrack reaching the artistic and technical standards required. This final mix must be suitable for the environment in which it is to be reproduced – the home, the cinema, the office, the airliner, etc. It may have to be compatible for both multi-channel stereo and mono reproduction.

Films mixed for distribution in the cinema can make full use of all the dramatic power that sound can bring; there are no distractions for the audience, little background noise and no neighbours to complain about the volume. Developments in digital sound recording and theatrical reproduction have allowed a final soundtrack to have a very wide dynamic range, but this is a dynamic range that is only suitable for the cinema environment.

The dynamic range recorded may vary from the sound of a gunshot to the silence of a grave. But to reproduce this range successfully in every environment, from mono home television set to digital surround sound 'first run' cinema, is not without compromise.

If a simple two-channel stereo mix for television is to be reproduced in mono, compromises may have to be made to the stereo image to allow for good mono reproduction. Sound effects will often need to be slightly reduced in level, so that they do not appear to be too close in the mono image. In addition, different sounds at either side of the stereo image, which may be comprehensible in stereo, may not be in mono when the image is combined to the centre, and may, in fact, sound too loud. This is further complicated when stereo pre-mixes are combined with further stereo pre-mixes. When using a stereo matrix recording system it is important that the mix is heard through the matrix, to ensure there are no phase problems. Artificial phase reverberation must be carefully controlled too – spectacular effects in stereo can easily become unintelligible when combined into mono.

Most viewers adjust their television sets to reproduce dialogue at an acceptable level, and with television mixing this is the starting point of the mix. Cinema mixing allows a much greater freedom, with the opportunity to almost lose dialogue under sound effects. Dynamic cinema mixes may, therefore, not be suitable for television transmission. Ideally a separate television mix is required.

Layback

In the re-recording process, the soundtrack is physically separated from the picture. Later, for motion picture theatrical release, it will be finally 'married' to the picture in the form of a multi-channel optical analogue or digital print. For video transmission, however, at some point the audio master soundtrack has to be 'laid back' onto the video master for transmission. Obviously, if the programme has been mixed and recorded onto the original edited master, this layback is unnecessary. Normally, however, the soundtrack is kept as a separate track from the picture in the audio post-production area, and has to be married back to the master picture. The synchronization of the layback is time code controlled. Since both picture and master soundtrack have the same code, a simple lockup with an audio transfer is all that is needed; this may or may not be produced from a virtual mix.

Music and effects tracks

Much of the material that is audio post-produced needs to be re-voiced into other languages for sales to foreign countries – for this a special soundtrack is made up, called an M & E (Music and Effects) track. It is sold as a 'foreign version' to countries that wish to post-synchronize their own language onto the material. The M & E track consists of all music and effects tracks, as well as vocal effects tracks such as screams, whistles, and crowd reactions tracks – in fact, sounds which cannot be identified as specific to a particular country. The balance of the mix should be as faithful a re-creation of the original language version as is economically possible. To assist in making a music and effects track, the final mix itself may be divided into and recorded as three separate tracks of dialogue, music and effects; these may be in stereo requiring six tracks. One ideal format to record these onto is the modular digital multitrack using removable discs or tapes.

In documentary material with dialogue and narration, the music and effects track often includes synchronous dialogue, but not narration. This narration will then be reread in the appropriate foreign language with the synchronous dialogue being subtitled or 'voiced over'. This saves the expense of post-synchronizing. Indeed, more and more foreign productions are now being sold with subtitles rather than post-synchronized dialogue. However, subtitles can often only summarize dialogue, particularly in fast wordy sequences. This means that the subtleties of the spoken word may be lost.

Creating M & Es is an additional expense which requires more sound mixing time, and often this cannot be justified unless the material is pre-sold to overseas countries. However, if a virtual mix of the project is held, re-creating an M&E is not time consuming, providing the tracks have been well laid and the dialogue is not laid in with the music and effects tracks.

17

Studio recording

It is often necessary to record dialogue for audio post-production, either as a narration (for a commercial or documentary) or when re-voicing foreign language films or post-synchronizing dialogue of poor recorded quality. Dialogue can be recorded in the mixing suite itself, although this is not ideal. There can be extraneous noises, the acoustics are not good, and headphones are needed to check the sound quality. It does, however, allow personal contact between the mixer and the artiste and can be satisfactory if a close microphone technique is used. Many small music studios work in this way.

In an ideal situation a separate studio is used as part of the audio suite, and this might be used for recording sound effects and music as well as dialogue.

Ideally, a studio used for speech recording should not suffer from low-frequency 'booming' and should have little or no self-resonance. The acoustic damping must not only reduce high-frequency reverberation, but also low-frequency as well.

The room should be insulated completely from external noise, and be equipped with the necessary facilities, including a time counter, talkback, foldback (i.e. a feed back from the sound mixing console), a cue light and a picture monitor or screen. In film operations the picture may be viewed directly via a film projector but more probably using a television monitor from a video transfer, since most film material is edited using a non-linear digital editing suite.

Recording dialogue

Performers delivering dialogue should be encouraged to stand so that the chest and diaphragm are unrestricted. However, to stand for long periods can be physically arduous, and many performers prefer to compromise by using a high stool and a lectern. Some prefer to sit at a table, although this may not be particularly satisfactory, both from the point of view of the performance and from the problems of sound reflections which a solid table top will induce. A fold-up music stand may prove a simple solution.

The quality of speech can be improved by equalizing the signal electronically through the recording console. Compression can raise average signal levels, thereby increasing intelligibility by reducing variations in level. A slow attack time is not advisable here, as there is a tendency for the high-frequency content of the sound to pass unattenuated which can lead to sibilance. Although equalizers specifically designed to reduce this are available, they are not particularly effective. Equalization can also help to correct over-emphasis of low frequencies occurring through close microphone techniques. A certain amount of mid-lift is often given to dialogue (between 2 and 6 kHz) to increase intelligibility, although this can again create sibilance if overused. Dialogue equalization is particularly useful when speech is to be replayed at high volume in a poor acoustic. Continuing improvements in loudspeaker design, recording quality and acoustic design has led to less need for dialogue equalization, particularly for films made for theatrical release – now they play smaller in theatres, with digital sound.

Post-synchronization

An original dialogue track may contain some material that is unusable; perhaps there is too much reverberation, too much background noise, a swear word needs to be removed, there is a fault in the recording machine or perhaps there was just insufficient time to set up the mics properly. When this occurs the dialogue can be replaced by a method known as post-synchronization.

The dialogue to be post-synchronized is divided into small sections of anything from two or three words to a sentence or two. These words and their appropriate pictures are then played back to the artiste in a studio. The artiste listens, and repeats the words in synchronization with the picture as he hears the original dialogue through his headphones. The words and pictures repeat themselves in the form of a regular loop: a cue appears just before the dialogue is about to start. The words to be spoken may be displayed below the screen, in synchronization with the picture. The scene repeats itself until the actor has achieved satisfactory synchronization, and a recording has been made.

Loop method

Originally using the film system, the sections to be re-voiced were made into two continuous loops of picture and sound. A streamer, or wipe, was physically drawn onto the picture with a chinagraph pencil, to cue the artist to start. When the actor had perfected his words, a recording was made and another set of loops was 'put up'.

This system has many drawbacks. Obviously it is tedious and time-consuming work to make up loops, and in large looping sessions storing loops becomes a problem. In the early 1960s an automatic dialogue replacement system was introduced to improve efficiency.

Automatic dialogue recording (ADR)

In this system, loops are made electronically. The ADR controller is given two points to shuttle between, and is programmed to produce a visual or aural

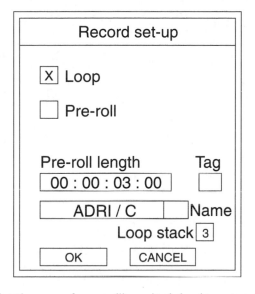

Figure 17.1 A workstation screen for controlling a simple looping system. Allowing nine separate takes of a loop, in–out points are set on a separate screen

cue of some kind just before the 'loop' is about to start. This is essential if the artiste is to start the loop accurately on time. The controller can be set to shuttle between any number of points, and after successfully recording one loop it is reset and moves on to the next loop.

The ADR system is well suited to digital disc recorders with their instant access, and workstations often offer ADR facilities (Figure 17.1). The number of loops that can be set and the number of 'recordings' possible on each loop vary between makes of workstations. But most are able to offer more than sufficient facilities – listening to 10 or more takes of one particular loop is something of a daunting task, making it very difficult to judge which is best.

The start of dialogue on a workstation can be cued by activating an event trigger using a GPI (general purpose interface). More sophisticated systems are available as part of Midi music scoring software. Here the software is driven by Midi time code and pre-rolls a clock to reach zero as a cue when a dialogue loop starts (Figure 17.2).

If only a small amount of ADR work is necessary, it may well be fitted into a final mixing session, or at a suitable time during the track laying. It is even possible for the work to be carried out in a portable studio at a location, which may be a necessity in a episodic production with a tight time schedule not allowing the performers to get away from the set. Complicated sessions need to be well organized so that artistes are only called in for a minimum of time, thus saving costs. The sound editor will produce a log of all the required lines by time from one performer. Looping is not carried out in order, but by the performer; in this busy situation, looping will carry on in parallel with sound editing. However, in most non-theatrical productions only a line or two will need replacing. Arguably, the better the sound recordist the fewer lines that need re-voicing!

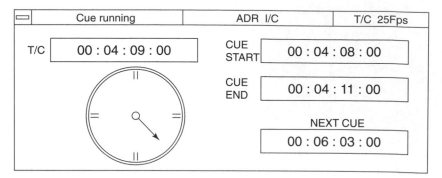

Figure 17.2 ADR clock cue

Figure 17.3 shows an extensive ADR screen on an audio workstation.

It is said that post-synching produces better performances from actors, since all their energies go towards giving a good verbal performance and they are not worried about remembering lines. But there is no continuity of recording and little interplay with fellow actors. Scenes do not flow and it is difficult to obtain convincing performances in a sterile sound studio.

Automatic dialogue replacement can be refined further by using automatic time alignment systems. Offered on some digital workstations, these compare

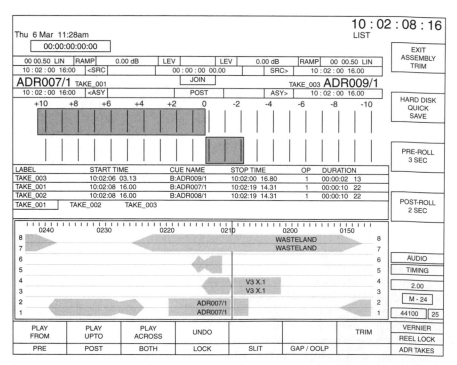

Figure 17.3 An extensive ADR screen on an audio workstation (*Courtesy AMS Neve Ltd*)

the relative timing difference between the post-synchronized dialogue being recorded, and the original unusable dialogue on the track. In ATA systems the post-sync dialogue is automatically edited within the machine to match the original dialogue. The system normally runs 'hands free' but it is possible for the operator to adjust the parameters of the system to tighten the 'fit' of the dialogue.

Generally speaking, post-synchronized dialogue is recorded in dead acoustics, with the necessary equalization and reverberation being added in the final mix when the overall effect needed can be judged.

Post synchronous sound effects (Foley effects)

When dialogue has been post-synchronized, all the ambient sounds of the original track such as footsteps, doors opening and closing, clothes' rustles, etc. are lost. However, these can be replaced, again by using post-synchronizing techniques (called Foley effects after Jack Foley who started the system in the USA). Often, these additional effects are required even when original synchronous dialogue is used, for the incidental effects may have been ignored in the location recording in favour of the dialogue.

To reproduce sound effects in a studio the scene is first viewed, and the actions are then mimicked using similar props to the original scene. Unlike dialogue replacement, however, longer sections can be recorded at one time and guide tracks are not necessary. Usually footsteps and clothes rustles are recorded in this way. Spot effects, however, are more difficult to synchronize accurately. Sometimes it is impossible to produce an exact sound effect from an actual prop because the real thing may not sound convincing; water spouting from a hose, for example, often sounds weak, but it can be transformed by adding the effect of a small waterfall. The sound of a large bird flapping its wings can be simulated by opening and closing an umbrella. Many contrived sounds are better than the real thing, and every possibility is worth investigating.

Soundtracks made specifically for foreign versions will often require to be post-synchronized with effects and footsteps since the dialogue will have been removed, complete with all the other 'live' sounds. However, often little time is available after a final mix to create complicated 'M & Es' and these may be rather cursory; much will depend on whether any foreign pre-sales have been made.

Narration recording

Not only are audio post-production studios used for post-synchronizing dialogue but they are also used for narration recording. Narrations, 'voice-overs' or commentaries in Britain can be recorded either before or after editing, but to assist the performer in giving the correct intonation it is better to record a narration to the edited picture. To record a narration without pictures, 'blind', may lead to a poor performance when later viewed against the visuals, although much will depend on the director. Often picture editing is quicker to a timed voice-over.

To prepare a narration, a shot list giving footages or time is usually made. It will give details of the visuals, with cues for effects and music, and the in and out times for the synchronous dialogue. The narration is then carefully written to the pictures, taking into account the reading speed of the performer, which on average can be timed by counting a second for every three words spoken (or in film, two words for every 35 mm foot that passes, or five words for every 16 mm foot). Valuable time can be taken up rewriting narration to fit. A single speed reverse motion control has its advantages here. If a voice-over overruns or is short, narration can be read starting at the end cue – the programme material is run in reverse – and backtimed; when the voice-over stops this will be the exact starting point for the run forward. Narration can be written to be informative or, alternatively, to be explanatory to the picture. Informative narration needs the viewer to listen with attention; explanatory narration tends merely to accompany the picture with relevant words; they add to the production in an entertaining way and simply become part of the whole.

Once a voice-over script is completed it is carefully marked with start and stop cues in the form of timings. In the studio, the performer begins reading when the first time cue appears and attempts to finish by the out-cue. Through headphones he or she will hear other relevant soundtracks which may provide valuable cue information – perhaps the narration introduces a performer who speaks on a cue. An alternative to using time cues is to cue by a light.

The voice-over script should be typed on only one side of the paper and laid out in front of the microphone so that it is unnecessary to turn pages. Should the narrator make a mistake in the middle of a piece, the recorded track can be run back and then rolled forward. The narrator will then hear his or her own voice through the headphones. The narrator can judge the type of intonation he or she was using and at a convenient point, prior to the mistake, the track is switched from 'replay' to 'record' and the performer picks up the reading. If the intonation is correct, the insert will be imperceptible.

ISDN

Sometimes it is impossible to arrange for a performer to attend a recording session because of the distance to be travelled. In this situation it is often cheaper to record the session in a local studio through a telephone line via an ISDN link. Through compression systems such as MPEG-Audio, good high-quality sound can be recorded; the system is simple to use and once the equipment has been purchased it is cheap to operate. Some voice-over artistes install their own ISDN links in their own home studios.

18

The transmission and reproduction of audio post-production material

Once audio post-production has been completed, the mixed soundtrack is ready for reproduction. This can be reproduced on a format of any quality, and may or may not be in a multi-channel format.

Low-quality formats

Low-quality formats include:

- Video cassette recorders (not hi-fi)
- Standard U-matic industrial formats
- Computer video games (at eight-bit quantization).

High-quality formats

High-quality formats include:

- Video cassettes with hi-fi sound systems
- Video discs
- Television transmission, using digital sound
- Digital film soundtracks
- Digital versatile discs.

Between these two quality extremes are:

- 35 mm optical soundtracks with noise reduction;
- Analogue sound television transmission systems.

In the cinema the sound-reproducing chain from final mix to released motion picture film is comparatively short and can be of the highest quality. In television the chain is longer.

The television chain

Replayed from the telecine machine or video recorder, audio post-produced sound and pictures for broadcasting are routed to the transmission area of the television station where they are monitored, equalized and then fed to their correct external destination. This area also routes tie lines, intercoms and control signals, and provides all the necessary equipment needed to generate pulses for scanning, coding and synchronizing video throughout the station. Some or all of the path may be in digital form, sampling at 48 kHz and quantizing at 20 bits.

Sound in stereo form is usually sent in a simple left/right format. The problems of routing stereo signals are very much greater than routing mono; not only must the signals maintain the same level left and right, but they must also maintain phase. Stereo matrixed signals from locally produced programmes and theatrically released feature films are also transmitted, enabling anyone with the correct equipment to pick up the surround sound. Monitoring the signal can create problems within a control room with limited space. Unfortunately, matrix signals suffer loss of stereo separation rather more easily than discrete left and right signals.

Any television organization using stereo must produce a decibel attenuation standard for the mono sum programme signals (produced from the left plus right stereo signals), and this figure must be standardized throughout the transmitting network. It is designated an M number. In the USA, M0 is used, in Europe M3 and M6. Stereo signals are usually monitored on left and right meters, with an additional third meter used to indicate the mono signals. More useful than a meter is the XY oscilloscope, which will show any instant out-of-phase condition. These are particularly important if the transmitting chain is long and complicated. In America, for example, a large network broadcasting company may have two network distribution centres, one on the east coast and one on the west coast, each distributing sound and pictures over satellite and terrestrial systems. These network centres can integrate programmes from videotape, film, remote pickups (outside broadcasts) and live studios. There are many pieces of equipment in the chain, including satellites, studio and transmitter links, processing devices, transformers, equalizers and much more. When the links are joined together, the effects of their individual degradations are cumulative and without proper metering or monitoring, the task of discovering the source of a problem can be enormous. The introduction of digital distribution reduced these problems dramatically with the added advantages of reduced maintenance and adjustment costs.

In transmission, a video signal leaves the studio centre by a microwave link or a coaxial line. The accompanying sound may be carried on a high-quality digital AES/EBU audio line, or sometimes within the video signal using a digital pulse code modulation technique, which inserts the audio into the line synchronizing pulses (sound in syncs). This has three major advantages:

1 It does not require a separate sound circuit and is therefore cheaper.
2 There is no possibility of the wrong picture source going with the wrong sound source.
3 Picture and sound stay in synchronization even if a delay is added to the source to synchronize with another incoming signal.

Transmission of television signals

Television signals can be transmitted into the home through cables, through the atmosphere from earthbound digital or analogue terrestrial transmitters, or through space, from satellites.

The signals transmitted through the atmosphere or space are essentially radiated frequencies vibrating above the frequency of the audio spectrum. It is only when current alternates at these so-called radio frequencies, of 20 kHz to 13 Mcs and beyond, that energy is thrown off and in such a way as to radiate and be received over distances.

These radio waves, like the signal generated around a wire carrying audio, are both electrical and magnetic. Their frequency distinguishes one from another. In a radio wave the electric and magnetic fields are at right angles to each other and to the direction of propagation. The speed of any radio wave is always the same whatever its frequency. Denoted by the letter 'C', it is equal to 186 000 miles per second, that is, the speed of light, which is the fastest speed known in the universe. Despite this speed, pictures bounced off the moon take over a second to arrive at the earth and even by live satellite, programme material takes over a quarter of a second to arrive.

The length in space of one cycle (or one vibration) is called the wavelength, and is measured in metres. This wavelength is spread over 'C' metres; therefore a radio station broadcasting on a frequency of 600 kHz has a wave length of the speed of the frequency of a radio wave in space (C) divided by its own frequency, which equals 500 metres.

It is now necessary to modulate the audio signals onto this radio frequency carrier wave. For television transmission the video signals are frequency modulated on the carrier, in a similar manner to the way video signals are recorded. The sound is transmitted on a slightly different frequency to that of vision, allowing sound and vision to be processed separately.

In the UK system the total channel width of the television signal is 8 MHz, the vision width is 5.5 MHz and the sound vision separation 6 MHz. Audio signals up to 15 kHz can be transmitted using analogue techniques.

The digital television system is shown in Figure 18.1.

Stereo television sound

Stereo sound for television can be transmitted in various ways, but the main concerns of any system must be to ensure that the stereo sound does not affect the quality of the mono transmission, and that the transmitted area in stereo is the same as for mono. In Britain the audio signal is spaced 6.552 MHz above the vision, and digitally modulated at a rate of 7.28 kbps. The resolution is achieved by using compansion to 14 bps, although it is transmitted at only 10 bps. It is called 'Nicam' (Near Instantaneous Companded). The format is similar to that of the MAC (Multiplexed Analogue Components) system which is used in European Direct Satellites. Satellites receive their signals from ground stations via microwave link. Arriving at the satellite, the signal is converted to the satellite's transmitting frequency and then sent back down to earth. In the European DBS system, the transmitting frequencies are in the range of 11.7 to 12.5 GHz with the carrier

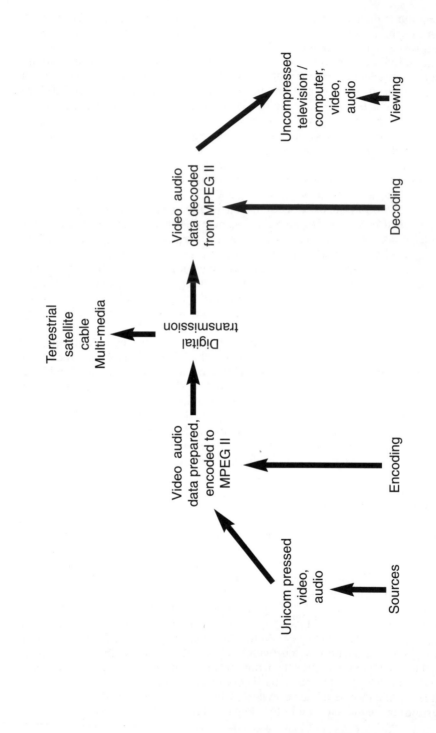

Figure 18.1 The digital television system

having a band width of 27 MHz, which is much higher than any terrestrial television system.

Any transmitted wave, from whatever source, has to contend with both absorption from the ground or the atmosphere, and reflection. Waves of low frequency are readily absorbed by the earth and objects in their path, and since losses increase with frequency, ground wave transmission is unsatisfactory over long distances.

Higher frequencies are, therefore, used in broadcasting, which take advantage of the ionosphere (layers of ionized air between 100 and 1000 kilometres above the earth). These layers can be used to bounce radio waves back to earth, like a radio mirror. However, these reflections back to earth begin to fail at radio frequencies of about 100 MHz, and completely disappear at about 2 gigahertz (GHz). At this point the waves travel straight through the ionosphere. The ionosphere is of no use whatsoever for mirroring back terrestrial television signals, the lowest of which is in the ultra-high-frequency range. Terrestrial television is therefore radiated over short distances, with many transmitting stations making up for the ground losses.

Satellites transmit at even higher frequencies than terrestrial television systems. They can project straight through the ionosphere with little interference (in the space above the ionosphere there are no transmitting losses), so transmission can be over very long distances. This means that one satellite can cover an area that would normally require several hundred terrestrial transmitters.

Whereas over 600 stations might be necessary to cover Britain using earthbound terrestrial television transmissions, one satellite can do the same job, with its footprint covering much of Europe as well, although there are obviously language problems and cultural barriers to overcome.

It was the introduction of compression techniques such as those specified by MPEG that made satellite transmissions cost-effective. A typical distribution satellite might be able to transmit 18 different analogue channels, but with digital processing for each analogue channel 30 individual 312-line VHS-quality channels can be sent or eight channels of standard analogue quality picture and sound. Digital television brings many more programme channels, home shopping, telebanking, data services for commercial and home use, and facilities for the purchase and downloading of computer software.

Surround sound

Dolby Surround is the consumer side of Dolby Stereo; any two-channel consumer format can be recorded with this matrix system and be decoded domestically. It is available in consumer equipment in both simple passive and active versions. The cheaper passive version, without amplifying circuitry, is licensed and marketed as Dolby Surround. In this domestic format the centre screen speaker is missing or is merely a sum of the left and right, which while reducing the hole-in-the-middle effect can also act to reduce the overall stereo effect. In Pro-logic Surround, the more effective active option, a discrete centre channel is offered, suitable for larger environments; it also offers reduced signal leakage between speakers and improved directionality. All systems offer a fixed time delay for the surround channel in the home environment; this will be between 25 and 30 milliseconds 5.1 Surround Sound. Discrete 5 channel audio is also available domestically, issued on video discs and DVD discs. It uses encoding and compression since traditional digital 16-bit 44.1 kHz audio takes up to much data in a five-channel format. Standard encoding systems include MPEG audio, DTS and Dolby Digital. Most widely known is the Dolby AC3 audio encoding system used in Dolby Digital, found in cable television, satellite television, digital terrestrial television (including high-definition television), digital versatile discs and ROMs, laser discs, PC sound cards, film soundtracks and so on.

This five-channel audio format uses a system called perpetual coding. Traditional 16-bit 44.1 kHz coding takes up too much data in a five-channel format and Dolby developed the first perceptual coding system that was intended specifically for multi-channel audio. The data rate reduction is achieved by adapting their established analogue-derived noise reduction system. Dolby noise reduction works by lowering noise when there is no signal but allowing strong noise to cover or mask the noise at other times – the noise now becomes imperceptible. Like its analogue counterpart, Digital Dolby divides the spectrum into narrow frequency bands; the coding bits are broken up and shared among these frequency bands, depending on various factors such as the frequency spectrum and the dynamic nature of the programme. These and other efficiency savings make AC3 coding an efficient way of delivering surround sound.

In an attempt to 'optimize sound for the particular listener', the decoder for Dolby's Digital multi-channel sound system modifies the sound output according to use. Within the system data it:

- identifies the programme's original production format, mono, stereo, etc.;
- optimizes the dialogue level, adjusting volume;
- produces a compatible downmix from the multichanel sources;
- controls dynamic range, particularly from wide-range cinema sound tracks;
- routes non-directional bass to subwoofers as required.

Domestic video cassette machines

Television programmes and films do not necessarily have to be transmitted into the home. They can be distributed by video cassette or disc and in fact over half the income from theatrically released feature films is derived from this type of domestic sale.

All domestic video formats provide stereo sound facilities. Originally, the audio track on the domestic video recorder was of very poor quality because of the narrow tracks and the very slow speeds. However, audio frequency modulation recording is now available on these formats, producing good-quality hi-fi sound.

Despite these vast improvements in video cassette sound quality, most consumers still listen on the linear tracks of non-hi-fi recorders. Those involved in audio post-production should be aware of the inadequacies of video cassette sound and the vast differences in sound quality between hi-fi and 'normal' formats.

The VHS (Video Home System) format

This was announced by JVC in 1976 and now holds the lead over its only serious competitor, the Beta format. The video recording system is similar to that used in the professional U-matic system. The VHS format uses two audio tracks of 0.35 mm in width, each with a guard band of 0.3 mm. There is no guard band between the audio track and the tape edge; this makes the format susceptible to problems of poor tape splitting. Recording pre-emphasis and de-emphasis is employed at a tape speed of 1.313 inches per second (3.3cm/sec). Dolby 'B' noise reduction is available on many machines. On the standard analogue tracks, an audio band width up to 10 kHz is possible. The VHS system is shown in Figure 18.2.

Beta format

Introduced by the Sony Corporation, Beta was first demonstrated in 1975. Three versions of the format have been produced. Only Beta II and III are

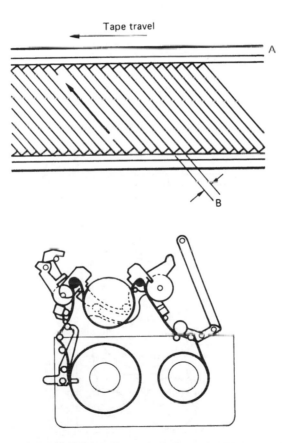

Tape travel

A

B

Figure 18.2 VHS tracks, tape width 12.65 mm. A, Audio track 1.00 mm; B, video track 0.049 mm; C, lacing pattern (Courtesy of *Sony Broadcast and Communications*)

now available. The sound is recorded longitudinally, running at a speed of 0.79 inches per second (2.0 cm/sec) and the two tracks are 1.05 mm in width. Frequencies up to 7 K can be recorded on the standard analogue tracks. The Betamax system is shown in Figure 18.3.

Beta Hi-Fi

The poor sound quality of the Beta format encouraged Sony to produce an improved Beta system. This was called Beta Hi-Fi. In the Hi-Fi format, audio frequency modulation recording is used with a noise reduction system, and low and high frequency pre-emphasis. The manufacturers claim an 80 dB signal to noise ratio, with harmonic distortion less than 1 per cent below clipping. Bandwidths extend to 20 kHz. This system is peculiar to America, for in Europe the PAL standard video recording spectrum does not have sufficient room for the insertion of the hi-fi carrier into the picture recording system, and a VHS Hi-Fi-type audio system is used.

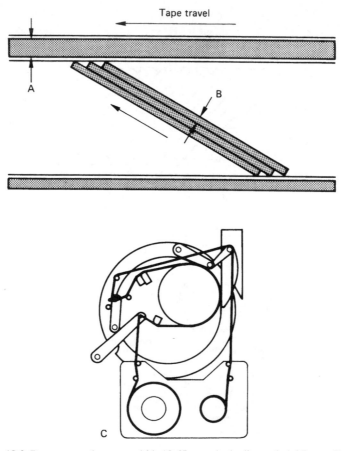

Figure 18.3 Betamax tracks, tape width 12.65 mm. A, Audio track 1.05 mm; B, video track 0.0328 mm; C, lacing pattern (Courtesy of *Sony Broadcast and Communications*)

VHS Hi-Fi

VHS followed quickly on the Beta Hi-Fi system with their own format. The VHS system modulates the tape using separate heads to those used in the video system, a compression and expansion system, and pre-emphasis and audio FM modulation. The audio is recorded below the video using a system called Depth Multiplex recording. Audio is first recorded on the tape from heads on the video drum. Video signals are then recorded over this, but only at the surface of the tape. It is possible for these two signals to be decoded separately later, in replay. VHS Hi-Fi is of similar quality to the Beta Hi-Fi format. Analogue tracks are also provided to create compatibility with standard VHS tapes.

Super VHS

Super VHS was launched in the USA in 1987 and Europe in 1988. It produces compatible pictures of a quality superior to standard VHS. The sound is recorded in a similar way to VHS Hi-Fi.

Video hi-fi formats are particularly sensitive to poor alignment. If the mechanical adjustments are imperfect there will be poor head switching and interruptions in the video signal. As a result, the audio will contain a low buzzing noise. This in turn creates problems in decoding the compression and expansion systems, particularly on low-level signals where the buzzing can even modulate the track.

The various standards that exist even within one format are even more of a problem. Minor variations can be found in the pre-emphasis used by different manufacturers and even by the same manufacturer.

DVD (digital versatile discs)

Introduced in 1997 in the USA and 1998 in Europe, this interactive audio visual disc produces high-quality multi-channel digital sound and vision. The audio is recorded in layers, allowing access via an interactive screen to various parts of the programme like a CD-ROM. Musical DVDs include biographies of the musicians. Drama-based discs give information about authors, locations and even offer alternative storylines. These are produced in the same way as any other programme material, but each section is coded to be called up as required.

The audio is encoded to reduce recording space. Various formats are recorded on to the DVD disc, and the player picks up the one appropriate to its systems.

Optical video discs

The optical video disc was first introduced in Europe in 1977. It is used mainly to play back motion picture films in the home. Picture quality is superior to that of the standard video cassette recorder and its sound quality is similar to any digital recording system. They are most popular in Japan, where the video cassette recorder has not gained the acceptance it has in Europe and America.

The discs are recorded and manufactured in a similar way to compact and DVD discs and the information is replayed by a laser beam pick-up device. Audio is available as an FM signal or as a PCM signal.

The basics of the computer and digital data recording

The digitizing of sound and video is a sophisticated and powerful modern technology which is currently revolutionizing the broadcasting and communications industries. Electronic digital technology arose half a century ago when it was discovered that calculating machines could still perform any sized calculation using just two digits, 'zero' and 'one'. This deceptively simple breakthrough enabled the transition from cumbersome mechanical calculators to lightning-fast electronic calculators and computers. The first electronic computer is reputedly Turing's successful code-cracker 'Colossus', dating from the second World War. This was followed in 1946 by the building in America of the even more colossal 30-ton, 17 000-valve Eniac computer, designed to be programmable for more or less any mathematical problem, not just code cracking.

All this gave the computer industry a huge mystique; computers were spoken of with awe and mistakenly attributed unlimited powers. However, the potential of digital computers was real, and today, by the integration of millions of separate components onto a few tiny silicon chips, computers have emerged from large, secret air-conditioned rooms into the hands and homes of us all. A humble word processor from the 1980s is faster and more powerful than the huge Eniac ever was.

The power and speed of electronic silicon chips has grown exponentially since the 1970s, but it is only comparatively recently that they have been able to meet the prodigious demands of the audio and video industries. The fact is that real-time video processing needs thousands of times more computer power than word processing.

At the heart of the modern computer is the *microprocessor*, which controls all the other devices within the machine. A microprocessor is a chip of silicon not much more than a centimetre square covered in millions of interconnected transistors. The chip has a couple of hundred connections to the outside world via fine gold wires to pins sticking out of the chip's protective package.

The chip's pins may all look the same but they are divided into different functional groups. For example, there is a 'data pin' group (usually 32 pins). The patterns of binary voltages (either 'hi's' or 'lo's' in electronics jargon) put onto these pins actually activate the processor's electronic circuits to do specific things. Since the patterns effectively control the microprocessor they are called 'instructions'.

Computer instructions and data are stored as lists in a computer's 'main' memory as computer programs. Each instruction tells the microprocessor where to get data, what to do with it, and where to get the next instruction. The microprocessor can therefore feed itself data and instructions indefinitely according to the wishes of the computer programmer who wrote the program.

The microprocessor chip has another important 32-pin group called the 'address pins'. The processor sets these to the address in memory that it wants to access. This is how the processor can automatically get its own next instruction after completing the last one. Modern processors can process instructions this way at rates of hundreds of millions a second.

The rest of the chip's pins enable the processor to control all the other chips inside the computer, and all the devices the computer uses – such as disc drives, monitors, keyboards, mice, printers, CDs and sound systems. Because of its controlling role the microprocessor is often also called the *central processing unit* (CPU).

Memory

Next to the CPU, memory is the most important component in a computer. It provides the CPU with instructions, it stores data bit patterns generated by the CPU, and it acts as a reference for many of the peripheral devices connected to a computer. Every memory location has its own unique address, therefore the system can allocate specific functions to certain addresses and then the CPU and all the peripherals know where to find the latest information on any particular function.

The block of allocated memory called the 'video memory' is actually what you see on the monitor screen. The computer's video circuitry scans the video memory 50 to 100 times a second and puts whatever it finds onto the computer monitor screen. In the early days each 'hi' or 'lo' 'bit' in the video memory simply determined whether a corresponding screen pixel (picture dot) was lit or was dark. This was adequate for text characters and line-graphics. Now video circuitry can interpret three bytes (that is, 24 bits of information per screen pixel) which can determine exactly which of 16 million hues and shades of colour each pixel should be. This gives true-colour pictures.

Central processor chips and video circuitry access memory millions of times a second, which means that the memory must be very fast and capable of completing an access cycle in a fraction of a microsecond. Ordinary *dynamic random access memory* chips (DRAM) are used for the bulk of computer memory and have an access time of around just one twentieth of a microsecond. This was adequate, once, for both main memory and video memory in computers running up to 20 MHz with low-resolution displays, but computers now have at least 200 MHz CPUs and 200 MHz band-width video. Therefore the video memory address space is filled with specially configured high-performance video random access memory (VRAM), and the main memory is buffered with a cache of high speed 'static' RAM (SRAM).

The central processor 'cache memory' is a relatively small block of the very fast and expensive SRAM which is better able to keep up with the processor's demands. The cache is kept loaded as far as is possible from the normal main memory with program instructions the CPU is most likely to need. Efficient caches are sophisticated and complex, and it has to be said, not infallible; they can actually slow down certain types of program.

Storage

Unfortunately, all main memory, caches and video RAMs go blank when a computer is switched off; they are, in fact, microscopic capacitors which are kept topped up by the power supply: turn off the power supply and they quickly discharge – losing the contents of the memory. Since all programs 'evaporate' when switched off, it is essential to be able to keep permanent copies of these programs. The copies are recorded in circular tracks on the coated surface of a spinning disc. The coating is magnetizable like the coatings on videotape. The whole recording and reading device is the computer's *hard disc*. The bit patterns are recorded in concentric rings by a read/write head on a very light, stiff arm which can swing accurately and quickly to any track, often in less than 10 milliseconds. This gives 'random access' to any bit of data on the entire disc. Since the disc can be spinning under the read-head at up to 5000 revolutions per minute, the data can be read off very rapidly at a speed of at least 10 megabytes a second. The disc itself is not scuffed by the head at this speed since it surfs at a micron or so above the surface and is actually sitting on the disc's boundary layer of air.

By stacking double-sided discs with ganged pickup arms, vast amounts of data can be stored in one hard drive smaller than a VHS video cassette. Capacities of 10 gigabytes (i.e. 10 000 million bytes) and more are available.

The user interface

In the early days of computers hardly anyone knew how to use them, and computer programmers, in many people's eyes, became the high priests of the scientific world. Working out what bit pattern sequence was needed to make the computer perform even the simplest task was an extremely difficult exercise, because the program bit patterns had to be put in by hand, each bit being turned on or off with banks of switches. These, the precursors of the microprocessor's pins, all had lights arranged in 'data'- and 'address'-type groups. They all flashed impressively on the front panel of the computer when it was running, and gave the programmer some idea of what was going on inside.

The bit pattern lists, i.e. the programs, were stored in arrays of magnetizable ferrite rings strung on wires by hand. This 'ferrite core memory' was therefore extremely expensive to make and was used with great economy by programmers. Hence if there were any error messages at all they were in very abbreviated jargon, often only code numbers, and incomprehensible to all but those in the know.

Now we have only to point and click a mouse to have a massively sophisticated program called the 'operating system' do our every bidding and guide us around problems with comprehensive messages. This is price related; now memory costs only a millionth of the price it was in the early days and program development 'tools' have come a long way from bit-switches and lights. Operating systems, even in the 1970s, could boast only a few kilobytes of memory; now they are a thousand times bigger, developed to serve the huge market for sophisticated operating systems which make computers accessible to everyone.

The nature of digital information

Computers effectively communicate with just one voltage which is simply either ON, or OFF (the 'hi' and 'lo' of electronics jargon). These two states can be clearly distinguished even in the presence of noise which would completely scramble any analogue signal.

It is easy to see how such a robust system could be the basis of an enormously reliable signalling system for a computing machine. It is not so easy to see how the finer things in life, such as hi-fi sound and moving pictures, could be reproduced with nothing more than ON and OFF signals.

The answer is to code meaning into strings of ON and OFF digits, and then play them at such huge rates that they can match audio and even video frequencies. Digital to analogue converters (and vice versa) accurately convert this digital information to analogue signals so that they can be heard and seen on conventional equipment such as speakers and monitors.

Bits and bytes

The fundamental unit of digital information is the 'bit'; it has only the two states, ON or OFF. These states are traditionally represented by mathematicians as '1' and '0', and referred to as 'Hi' and 'Lo' by electronic engineers. The two states can be physically represented in a wide variety of ways from voltage peaks in wires to pits in plastic (for example, compact discs). Digital sound and pictures can be transferred from wires to magnetic tape, to plastic CD, and back again with perfect fidelity, providing they all run at the same speed.

The principle of coding meaning into digital strings of data is shown in the SMPTE/EBU time code structure used to synchronize film and video frames and sound recordings electronically.

Traditionally, strings of digits are broken up into groups of eight, called 'bytes'. The meaning of digital strings comes from the patterns of 0s and 1s in each byte. There are actually 256 different patterns possible so a byte can 'mean' 256 different things, usually just the numbers 0 to 255; but in word processing, for example, bytes can represent all the letters and symbols on a keyboard. For audio and video the bytes can represent 256 different levels of sound for audio, and 256 levels of light for video.

Analogue to digital conversion

The whole digital revolution in communications, audio and video is dependent on devices called analogue to digital converters (often abbreviated as

'A / D converters'). They are hybrid integrated circuits designed to take in an analogue signal at one end and produce the digital version from the other.

They work on the principle of sampling the analogue signal voltage and generating a digital number proportional to the voltage at the sampling time. They do this very quickly indeed, so quickly that they can follow the rise and fall of video frequency wave forms.

The quality of an A/D converter is judged not only by the rate of its sampling frequency but also by its resolution. An eight-bit A/D converter can resolve signals into just 256 levels, whereas a 16-bit converter can resolve 65 356 different levels and is hugely superior.

Once a signal is converted from wavering analogue voltages to a set of definite figures, it can be stored and copied exactly. And the figures are amenable to mathematical and logical manipulation, for example by frequency spectrum analysis followed by digital filtering and boosting. For audio post-production this gives very fine control of the final sound quality.

For video, sophisticated image processing becomes possible, not only to improve quality but also to create a wide range of effects to enhance the impact of the image. These digital video effects can make things happen which might never otherwise be witnessed – dinosaurs can roam about nibbling the tops off tall trees and men can fly!

Digital to analogue conversion

Having produced sound or video in digital form and produced all the necessary processing, the digital data has to be converted back into an analogue signal in order to hear it on speakers and see it on a screen.

Back in the 1970s, home computer enthusiasts made their own digital to analogue converters. They just connected eight specially chosen resistors to the output pins of their eight-bit microcomputers. The resistors actually halved in value from pin to pin, so that, for example, if the first pin delivered one microamp when it was 'hi', the second pin would deliver twice the current, that is, two microamps. The third pin would deliver twice this – four microamps, and so on up to the eighth pin which would deliver 128 microamps (1 doubled seven times is 128).

All the resistor outputs were connected together and fed directly to a small speaker. Then when the byte values ranging from 0 to 255 were sent to the microcomputer output port, the fluctuating current in the speaker created sound. Ingenious programming could create the notes of the musical scale, and control their volume. The result was the characteristic 'buzzy' computer music, but it was real music to the amateur programmer's ears. To this day all PCs have a similar, crude digital to analogue converter and a small speaker to produce the computer's well-known 'beeps'.

Digital sound quality

Modern digital computer sound systems now deliver not just one byte at a time to a speaker but two or more 44 000 times a second, giving excellent dynamic range and level resolution. All this requires high bit rates, in the region of a million bits a second. Such rates are within the range of home computers which can process and move bytes easily at several millions a second. But all those bytes have to be stored, and even our 600 MB audio CD only holds about an hour of digital hi-fi sound.

Digital video

What really challenges the home computer is handling moving pictures. The rate needed to shift bits to digitally re-create broadcast quality TV pictures is awesome.

A single frame of UK broadcast TV is equivalent to a 768×576 pixel computer picture and requires three bytes per pixel to define the exact colours. This means about 1.3 MB per frame and there are 25 frames displayed every second which would require a total bit rate of 265 million bits a second (about 33 Megabytes a second)! For the 16:9 wide-screen HDTV standard with 1920×1035 pixel frames and 30 frames per second, a computer needs to shift 1.4 Gigabits a second (almost 180 MB a second)! Some home computers can shift data at these rates, but real-time image processing or image data generation at such speeds is extremely demanding.

The storage of digital video is also a challenge when you need 33 MB of space for each second of broadcast TV. In fact these data volumes are difficult to manipulate in a computer, especially in operations like editing, and the industry has had to resort to data compression. However, digital data, being just numbers, lends itself to precise mathematical and logical compression and decompression. (This does not entirely solve the editing problem though, as we will see later.)

Digital compression

Broadcasters compress digital TV picture colour and luminance information into two bytes per pixel, which reduces the broadcasting data rate by a third. However, there are aspects of moving pictures which make them amenable to much greater compression than this. For example, an area of black in a frame would result in long strings of zero bytes in the digital representation. Instead of storing all these strings of zero bytes they can be replaced by a code which simply says how many repeat bytes there are in each string. This is called RLE compression (Run Length Encoding).

Backgrounds in moving pictures often change very little from frame to frame, therefore a background only need be recorded once in detail, followed by just the small changes which occur from frame to frame. This is called 'inter-frame' compression. Other techniques used include what could be called 'wallpaper' compression. The computer finds the wallpaper's repeating motif types, and allocates each type a short identity code. The motifs only have to be recorded once, then just a few motif codes and motif coordinates can be used to record any area of wallpaper.

On decompression these three techniques can give precise reconstructions of the original. There are also ways of removing information that the eye and ear do not notice. Such compression techniques are appropriately called 'lossy'.

All these basic principles can be elaborated on and combined to give a whole range of very sophisticated compression systems. The degree of digital data compression possible for video sequences can be as low as 2:1 and as high as 100:1 for which it has to be said the resulting picture quality is poor; 12:1 is the maximum used in broadcasting.

Compression takes advantage of the eye's sensitivity; it cannot easily gauge absolute colour hue, but can gauge luminance. Therefore compression systems only remove the minimum of luminance, but remove up to 95 per cent of the colour information (i.e. precise Red:Green:Blue ratios).

Among the professional compression standard systems are those from MPEG (The Moving Picture Expert Group). These have been developed from the popular JPEG (Joint Photographic Expert Group) compression standard originally used for still colour pictures on computers, but later extended for moving pictures. MPEG is actually a 'lossy' compression system since the original is not perfectly reconstructed on decompression. The amount of loss can, however, be adjusted at compression to suit the application, depending on whether quality or storage is the main criterion.

The format MPEG II is used by broadcasters as it can give a compression ratio of about 10:1 while maintaining broadcast quality on decompression. The compression system is very processor intensive and is only possible in real time on home computers when they are supplemented with special MPEG compression hardware.

Editing

MPEG II (also known as ISO/IEC-13818) is an 'inter-frame' compression system which means that not every frame of compressed data will decompress into its complete original frame; it may well only contain information on differences relative to a previous frame. This means examining anything up to a dozen frames next to a chosen frame (to find a complete 'key frame') before one can reconstruct the original chosen frame.

This is a bit of an irony since compression was devised to make vast quantities of digital video information more manageable; now when editing to frame accuracy there is some analysis of other frames needed before reconstructing one particular frame.

MPEG is an agreed standard and has been adopted for broadcast and studio productions; here editing and image processing must be performed with no visible degradation of the image. Other uses include encoding both pictures and sound for cable and satellite transmissions and Digital Versatile Discs; here there are competitors to the MPEG system.

Disc storage

One of the aims of data compression systems is to get reasonable amounts of digital sound and video data on and off computer hard discs at standard frame rates. As time goes on, digital technology will inexorably advance; hard discs will become faster and faster, their storage will be measured in terabytes, and processor speeds in gigahertz, and then the need for compression and its attendant quality compromises and editing complications will be regarded as 'an old twentieth-century problem' – rather like the need for stable mains power to synchronize early film systems.

Digital buffers

Once compressed, video and sound data streams are no longer relative to real time. Playing a compressed digital video tape at constant speed gives bursts and lulls of frames, and in some formats these are punctuated by intermittent bursts of sound digits. This erratic output obviously cannot go directly to the screen and speakers, so it is split into two separate solid state memory buffers which smooth out the flow of frames and sound. The buffer memory can be dual-ported for speed; one an input port for the fits and starts of information coming from the decompressor, the other an output port feeding out a steady stream of video frames or real-time sound.

Digital stores – the future

Solid state memory chips have millions of memory cells etched on their top surface. At least four broadcast quality colour TV frames can be stored on the surface of just one chip with an area of only about one square centimetre. This is amazing but not quite amazing enough to be very useful – to record, say, half an hour of video will require not just four frames but about 40 000 frames of pictures, equivalent to a very large area of chip surface.

Thus the future for solid state memory is not in covering chip surfaces but in filling three-dimensional volumes with memory cells. Imagine many micron-thin layers of memory cells stacked a centimetre deep. This could give 10 000 layers of memory cells providing storage for a half-hour video's 40 000 frames in a memory no bigger than a sugar cube.

Further, the potential storage capacity of a centimetre cube of material using molecular-sized memory cells is in excess of 24 hours of broadcast quality video without any need for compression. These molecular video cubes will perhaps be accessed by shifting the frames up to the top face, where they can be read, while also copying them back down onto the vacant bottom face to preserve them. Using parallel optical reading and writing, the spooling process could easily run through a million frames a second. This would give random access within 20 milliseconds to any frame in a half-hour video (much the same access speed as a hard disc).

If such video cubes become possible, they may perhaps replace video tapes in cameras and VCRs (VCR might then stand for video cube recorder). This is the future one manufacturer sees – the development of recording not to mechanical tape cassettes or computer hard discs, but to solid state memory with no moving parts at all.

Film editing

Little film editing is now carried out, but this appendix provides an outline of the process.

In film editing the picture and sound are held as separate rolls of stock rather than as one combined picture and sound tape.

Film picture editing

The main item to be found in a film editing room is the editing machine. This device allows the picture to be viewed at variable speeds from still frame up to a fast speed of six times normal and more. In addition, one, two or three sprocketed magnetic soundtracks can be interlocked with the picture and driven synchronously. (See Figure 7.2.)

After a film has been shot on location, a copy or print of the camera negative is made and the location sound is transferred to sprocketed magnetic film. In the editing room the sound is synchronized with the picture. The clapperboard provides sound and picture information for matching the start of the shot. The picture is held in the editing machine's gate, at the point where the clapper is seen to hit the board, and the sound is held at the point where the clap is heard.

These two are now synchronized and run together. After the first take of the sound and picture have been synchronized, the editor runs the picture to the next clapperboard and joins the second piece of magnetic film to the first. This track is adjusted so that the second soundtrack runs in synchronization with the picture.

When the whole synching-up process is completed, the picture and sound may be run through a numbering machine which will print, onto every foot of the picture and its appropriate soundtrack, a unique alphanumeric code that will confirm synchronization between the image and the soundtrack.

The picture is now ready for editing, and the unwanted picture and soundtrack are cut out. Some of the sound may be stored away and later cut back into the soundtrack. The picture is now ready for spotting.

Film track laying

The edited picture and its appropriate soundtracks are now placed in a synchronizer for sound editing. This consists of a picture viewer, connected to a series of large sprockets. These sprockets are locked on a common drive shaft which can be moved by a hand crank; a footage counter is also provided. Each of the sound sprockets, normally three, is fitted with a replay sound head connected to a small mixer. The various soundtracks are built up on this device.

Soundtracks laid against a picture (schematic layout)

Numbered leaders are placed on each of the soundtrack wheels of the synchronizer, the soundtrack leaders being in synchronization with each other and the picture leader. The synchronizer is then wound down to the first frame of picture. At the first frame of picture, there is perhaps a scene in a city, with the sound of traffic, the effect of a distant street vendor and a car door opening. This scene was shot mute. The appropriate rolls of magnetic film, ordered from the sound effects library, are now laid against the picture.

Since a film cutting-room does not normally include any recording equipment, soundtracks that are needed for editing have to be ordered from a transfer house. The soundtracks are usually recorded to peak level irrespective of the volume at which they will finally be played. This keeps background noise to a minimum.

Sound editing starts at the first frame of picture on track 1. Magnetic film is spliced physically to the leader, the synchronizer winding from left to right. At this point, other tracks on the synchronizer will run off, since there is no sprocketed material after the numbered countdown head or leader. The effect for track 2 is not required for another 30 seconds, and the effect for track 3 for another 45 seconds, so additional sprocketed stock called a 'leader', or 'spacer', with no oxide or emulsion on it is spliced into these tracks on the left-hand side of the synchronizer. As the synchronizer crank is turned, the soundtracks on the right-hand side remain in synchronization, together with the picture. Next, the shot of the distant street vendor is found (it is not synchronous) and the appropriate sound is cut in at approximately the right time. This is then examined to check synchronization; by listening and watching it is decided that the soundtrack is about 10 frames too early. Therefore 10 frames of spacer are cut into this soundtrack at the left-hand side of the synchronizer, between it and the sound effect. This means that the street vendor will be heard 10 frames later. Similarly, frames could be removed at this point so that sounds are heard earlier; 'atmos' tracks and effects are now being run together, one atmos on track 1, with a separate effect on track 2.

At this point in the film we cut to a car door slamming nearby, and a character walking down the street. This was also shot mute. Our original atmosphere track is suitable for this scene since it tells the viewer that this scene is at the same location, but now we also need spot effects of a car door slamming and also footsteps.

To lay spot effects in perfect synchronization, the picture must first be carefully marked up. The exact point where the spot effect occurs is found on the picture and a mark is drawn, with a chinagraph pencil, where the car door is opened. The appropriate pieces of sound are now placed in the synchronizer and the exact points where these effects are heard on the sound stock are marked with the chinagraph. A written identification is sometimes placed on the film. The magnetic track is now placed in the synchronizer, with the chinagraph mark on the picture and the chinagraph mark on the soundtrack matched. The track is now cut into the sprocketed spacer or leader, to keep it in synchronization with the picture.

At this point the dialogue starts and is off-laid as necessary.

If the track laying is complex, the various tracks will each be categorized, as in any track-laying operation, for a specific type of sound. A logical layout of the tracks will help in making any final mix-down smooth and speedy.

Loops

Sometimes there may be insufficient time to prepare a specific atmosphere track, in which case a loop of the effect may be taken to the mixing theatre where, if necessary, it can be run in synchronization with the picture.

Cue sheets

The final operation, prior to mixing the film, is to make up a cue sheet to assist the mixer during the mixing session. The cue sheet is usually prepared in the synchronizer. Before starting to prepare a cue sheet or chart, the counter on the synchronizer is zeroed to the first frame of picture after the head leader. The synchronizer then counts up frames and feet in 35 mm film, 16 mm film or time, as the crank is turned.

Glossary

Access time   The time taken to retrieve video or sound information from a store, which might be in the form of a video or audio tape recorder, a sound-effects library etc.

A/D converter   A device which converts analogue signals to digital.

Address   A time or location within programme material; often selected as a 'go to' for machines via time code.

ADAC   Adaptive Transform Acoustic Coding MiniDisc encoding system.

AES/EBU   American Engineering Society/European Broadcasting Union. A standards body. Colloquially used to refer to the digital stereo interface AES 3.

AFM   Audio frequency modulation. See *Frequency modulation*.

Ambience   See *Presence*.

Amplitude   The maximum height of a waveform at any given point.

Analogue (audio and video)   A signal which is a direct physical representation of the waveform, in, for example, magnet recording or optical recording.

ANSI   American National Standards Institute.

Apple Macintosh   Computer manufacturer. A system favoured by some designers and graphic artists. The basis of some audio workstations.

Assembly A mode   Automatic sequential assembly of edits from an EDL.

Assembly B mode   Automatic assembly of edits nonsequentially in checkerboard fashion from an EDL.

Assembly   The first process of editing, when the various shots are joined together in a rough order to produce a rough cut. In linear video editing further edits mean additional copying that may be unacceptable from a quality point of view.

Assembly editing   Applied to linear video editing by adding one shot after another, each shot using its own control track. Therefore a picture roll may take place at the edit. See *Insert editing*.

ATM   Asynchronous transfer mode.

Attenuate   To reduce the amplitude or intensity of a sound.

Auto assembly   See *Conforming*.

Back time   The calculation of a start point by first finding the finish point and subtracting the duration. This can relate to edit points or the recording of narration.

Backup   A copy of a disc or file in case it becomes corrupted.

Balanced line   A method of sending a signal via two wires, neither of which are directly tied to the earth of the system. If an external electrical field is induced into the circuit leads, it is in phase in both leads simultaneously. This unwanted signal will be eliminated by cancellation at the balanced input point, while the audio signal passes through, unaffected. Applies to both analogue and digital lines.

Barney/Blimp   Camera cover to reduce mechanical noise.

Baud   Unit for measuring the rate of digital data transmission.

Bel   A relative measure of sound intensity or volume. It expresses the ratio of one sound to another and is calculated as a logarithm. (A decibel is a tenth of a Bel.)

Betacam (SP)   A Sony trademark, a professional analogue half-inch video format; SP means superior performance. The standard multi-use format of the 1990s.

Bias   A high-frequency alternating current (up to 120 kHz) fed into an analogue magnetic recording circuit to assist the recording process.

Bit   A unit of binary code, either one or zero. Represented in digital electronic circuits as a 'hi' or 'low' voltage.

Black burst   Blacking; colour black; edit black. Provides synchronizing signals for a system to lock onto for stabilizing a videotape recorder.

Blanking level   The zero signal level in a video signal. Below this is sync information, above this visible picture.

Blimp   A soundproof cover to reduce the mechanical noise from a film camera.

Bloop   An opaque patch painted over an optical soundtrack join to eliminate a click on reproduction.

Bouncing   Recording, mixing, replaying and then recording again within the audio tracks of an audio or video recorder.

Buffer   A RAM device used in digital sound processing (DSP). (See RAM.)

Burnt-in time code (window dub, USA)   Time code transferred with picture from the time code track of a videotape and visually displayed in the picture as part of the image. This can only be used on copies for audio post-production use. The burnt-in time code is accurate in still frame, whereas the time code figure from the longitudinal coded track is invariably wrong.

C format   A one-inch tape, reel-to-reel, video recording format using analogue recording techniques developed by Sony and Ampex. The standard format of the 1980s.

Camcorder   A combined video camera and recorder. Often similar in size and facilities to a film camera.

Capacitance   A measurement of the electrical storing ability of a capacitor or condenser measured in farads.

Cathode ray tube (CRT)   The picture tube of a television monitor, visual display unit, or phase meter.

Clapperboard   A pair of hinged boards which are banged together at the beginning of a film double system sound shoot, to help synchronize sound and picture; it may include time code information as well as the normal identification marks.

Clash   The term used to describe the point at which the two ribbons of a light valve come into contact and produce heavy distortion.

Clean FX   Effects which have no additional unwanted sounds.

Click track   A sound with a regular beat used for timing in music scoring, now produced electronically, but once made by punching holes at regular intervals in film and reproducing this through an optical sound head.

Client   Local area network receiving device.

CMX   Computer editing system, famed for their industry standard EDL software.

Colour bars   Electronically generated set of video reference signals consisting of vertical bars of red, green, blue, magenta, yellow, cyan, grey, white and black. Usually found at the start of a video project.

Colour framing   In PAL the sequence of 8 fields (4 frames) which comprises the repetition rate of the matching set of fields. (NTSC 4 fields: 2 frames) with videotape composite recording. The tape is only correctly edited if field 8 is followed by field 1. If this does not take place, a flash or jump will accompany the edit.

Combined print   A positive film on which picture and sound have both been printed.

Compression   A method of reducing the volume range of sound. Also, a method of reducing the storage requirements of video or audio during conversion to digital.

Conforming   The matching or copying of one source such as audio to another according to an EDL based on time code information.

Console   Colloquial term for an audio mixing device, called a 'desk' in Britain.

Control track   A regular pulse that is generated during videotape recording for synchronizing; it provides a reference for tracking control and tape speed.

CPU   Central processing unit, the main brain of a computer.

Crash   A systems failure in a computer.

Crosstalk   In stereo equipment a crosstalk occurs when some of the left-hand signal leaks into the right-hand signal and vice versa. Time code signals can also crosstalk into adjacent analogue audio tracks. Digital systems are free of crosstalk.

Crystal lock   A system of driving a video or audio tape recorder at a known speed, at high accuracy; accuracy depends on use and cost, ranging from one frame in 10 minutes to one frame per day or greater. The timing synchronizing signal is derived from an oscillating quartz crystal.

Cut away   A shot other than the main action added to a scene, often used to avoid a jump cut or to bridge time.

Dailies   See *Rushes*.

dB (decibel)   A convenient logarithmic measure for voltage ratios and sound levels. One tenth of a Bel, a 1 dB change is a very small but perceptible change in loudness. A 3 dB change is equal to double the power of the signal. A 6 dB change is the equivalent to doubling the perceived loudness.

DAT   Digital Audio Tape.

DAW   Digital Audio Workstation.

DDP   Disc Data Protocol (for CD mastering).

De-esser   A device for reducing sibilant distortion.

D format   D1, D2, D3 and D5 digital video recording formats. D1 and D2 are most common.

Digital (audio and video)   A signal where the waveform is encoded into binary form for storage or processing. This signal can be copied repeatedly without degradation (compare with analogue).

Digitize To convert analogue pictures or sound to digital.

Discontinuous shooting Shooting video or film with one camera and moving the camera position for each change of shot.

Distortion An unwanted change in a signal – the difference between the original sound and that reproduced by the recording machine. Distortion takes many forms, frequency, phase or non-linear. Non-linear distortion is usually measured as harmonic or intermodulation distortion and has an 'edgy' sound. Digital recording offers the maximum potential for distortion-free recording.

Dolby Manufacturer of various sound processing devices, noise reduction systems, optical sound systems, and digital encoding systems.

Double system recording Double system or separate sound recording. This is a production method which employs separate machinery to record sound and picture – it is used in film sound recording.

Drop frame American system of time code generation which adjusts the generated data to compensate for the speed of the NTSC television system running at 29.97 frames per second.

Dropout A momentary reduction or loss of signal.

DSP Digital Signal Processing. Integral to computer data processing of audio and video signals.

Dubber A playback-only transport used in film re-recording driven by bi phase signals; may be sprocketed film or modular digital multitrack.

Dubbing mixer British term for re-recording mixer.

DVE Digital Video Effects.

Edited master The final edited project with continuous programme material and time code.

Edit points, edit in, edit out The beginning and end points of an edit when a video programme is being assembled or soundtrack being recorded.

Edge key numbers/code The latent image of alphanumeric numbers recorded on the side of film to identify each frame after processing.

EDL Edit Decision List.

Effects Sound effects, FX for short.

Ethernet A form of local area network (LAN) using coaxial cable, connecting up to 1000 devices within 1.25 mile radius.

Equalization (eq.) The boosting or decreasing of the intensity of low, middle or high frequencies to change the 'sound' of programme material. Equalization is also used in analogue tape machines to overcome losses in the recording processes. These equalization characteristics are not standard; they are set in the USA by the NAB and in Europe by the IEC.

Event number A number assigned by an editor to each edit in an edit decision list.

Exbyte tape A data storage tape using the original Video 8 format adapted to data storage, sometimes used for transferring data to and from a workstation.

Exciter lamp The lamp in an optical sound reproducing machine whose light is focused onto a photo-electric cell. It is the interruption of this light by the varying density of the film that causes variations of the output of the cell and this amplified produces sound.

File   A package of digital data used in computer systems.

File header   Information regarding a file held at the start of the data.

FDDI   Fibre Data Distribution Interface (fibre optic).

Floppy disc   Flexible magnetic computer disc of 3.5″ (and 8″ or 5.25″ – now redundant).

Flying spot scanner   A type of telecine machine.

Foley (USA)   Creating sound effects by watching the picture and duplicating the actions.

Frame   One film picture, one complete video scanning cycle.

Free run   Applied in time code to the recording of time code from a generator that runs continuously. A recording therefore has discontinuous time code at each stop and start. See also *Record run*.

Freeze frame   The showing of one single frame of a videotape or film.

Frequency   The number of times of an occurrence in a given time, e.g. frequency per second, in sound (called Hertz). (See Figure G.1.)

Frequency modulation   A method of encoding a signal whereby the programme material varies the frequency of a carrier signal – the level of the carrier remaining constant. Audio examples include stereo VHF radio and the hi-fi tracks of some domestic video cassette recorders and in the Betacam SP and M broadcast formats. AFM recording gives excellent signal to noise ratio and frequency response. In video recorders it is recorded on the spinning head drum resulting in low 'wow' and 'flutter', which contributes to the good dynamic range.

Full coat/stripe (fully coated GB)   Full(y) coat(ed) film stock has oxide across the entire width of the film. Stripe has the oxide just in the area where the recording takes place.

FX   Short for sound effects.

Gain   The extent to which an amplifier is able to increase the amplitude of a given signal; the ratio of input to the output level.

Gamma   The degree of contrast attained in a photographic image. In optical recording it needs to be at a maximum.

Gen lock   A system whereby all video and digital equipment is locked to a common synchronizing generator allowing it all to be synchronized together for editing etc. It is vital that time code is in synchronization with the video if computer-controlled editing is to be used.

Gigabyte   A unit of computer memory capacity equal to 1024 megabytes (MB).

Guide track   A soundtrack made to help post-synchronization, and not intended to be part of the final mix.

Haas effect   A psychoacoustic phenomenon where the mind identifies the first place from which a sound is heard as the origin of the sound, ignoring other sources of sound arriving a fraction of a second later. (Used in Dolby Stereo.)

Hard disk (HD)   A data storage and retrieval device using magnetic discs and having random access.

Hardware   The physical components that form part of an audio post-production system – for example, the tape recorders, video player, and computer.

Header (USA)   A leader at the beginning of a programme.

Head out   Tape is stored on the reel with the programme material head out – recommended for videotape storage, rather than audio tape.

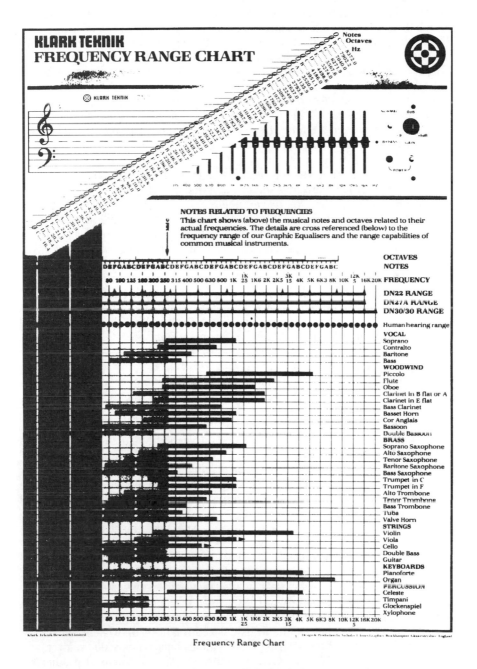

Figure G.1 Frequency range chart (*Courtesy* of *Klark-Teknik Research Ltd*)

Headroom   The amount of signal overload that can be recorded above the specified reference level without distortion.

Hexadecimal display   A facility for displaying 16 discrete characters 0–9 and A–F, as found in a time code display.

Hi-fi (high fidelity)   A term coined for the consumer market to describe state of the art audio quality and equipment.

House sync   A central timing reference used to synchronize transports and digital equipments. See also *Gen Lock*.

IEC   International Electronic Committee, a European Standards body.

Impedance   The opposition to the passage of electric current by the resistance and reactance of a circuit. In a tape head wound with many coils of fine wire, there is a high impedance. A low impedance head is wound with fewer turns of thicker wire.

Inductance   A device which provides a magnetic opposition to any growth or decay of electrical current – such as a choke or coil.

Ink numbers (US)   See *Rubber numbers*.

'In point'   The beginning of an edit, the first frame to be recorded.

ISDN   Integrated Services Digital Network, a system that uses digital data for high-quality audio and picture transmission on telephone lines etc.

Insert editing   Linear video editing where control track, time code, and black level have been already recorded on the tape (pre-striped); video picture material is added as the edit progresses (as opposed to assembly editing).

Intelligent synchronizer   A microcomputer that monitors machines in audio post-production situations and learns their ballistics and operational characteristics in order to speed up synchronization.

Interlock   A term to indicate that two or more pieces of equipment are running in synchronization.

Intermodulation   The result of the interaction of two or more frequencies – the amplitude of one is modulated at the frequency of another.

I/O   inputs/outputs.

ISO   International Standards Organization.

Jam sync   A process of locking a time code generator to an existing coded tape so as to extend or replace the code. Used when code is of poor quality.

Jaz   A make of removable disc drive.

Jog   A function on a video or audio tape recorder which allows tape to be moved at any speed through stop to single speed with pictures forward and reverse. (See *Shuttle*.)

JPEG   Joint Photographers Experts Group. An international standard for file compression that is used to save storage space for photographic images and frames of video in NLE systems. The levels of compression affect the quality.

Jump cut   A cut which breaks continuity.

L Cut   See *Split edit*.

Laptop   A small computer/word processor often used in the field to log shots, but can also refer to small video and/or sound editing computers.

Laying sound   To place sound in its desired relationship to picture.

Layback   Process of transferring mixed audio back to its video master.

Laydown, GB Layoff   Transferring source audio onto a post-production medium.

LAN   Local Area Network used to transfer digital data from a client to a server, for example from one workstation to another.

Leader In America identified as a leader or trail leader. The leader gives a countdown to programme start, in film using flashing numbers, in video a clock. See *Run-up/run-out*.

Level Volume of a sound. See also *VU meter, Peak programme meter*.

Level cut, straight cut (Editorial cut, USA) A point where sound and picture cut at the same point.

Level synchronization A point where picture and soundtracks are in alignment. USA term, editorial sync.

Longitudinal time code (LTC) Time code information recorded as an audio signal on a VTR or audio tape recorder.

Loop A loop of material running sound continuously. It may be an NAB cartridge, a digital sampler, or a piece of sprocketed film. The term looping refers to post-synchronization.

Lossless compression Process of compressing sound or picture without losing any data, particularly for transmission and recording.

Lossy compression Process of compressing sound or picture which results in some loss of data, particularly for transmission or recording of sound and picture.

LTC Longitudinal time code (often called SMPTE).

MB Megabytes. A measure of computer storage capacity equal to 1 048 576 bytes or 1024 kilobytes.

Macro A sequence of operations stored by the user under one key.

M & E A music and effects soundtrack that includes information that is only relevant to international sales, i.e. there is no dialogue.

MII format A trademark of Matsushita/Panasonic. A professional half-inch video cassette format.

MADI Multi-channel Audio Digital Interface capable of carrying 56 channels. Also knwon as AES 10.

MC Midi clock.

Mag track Sprocketed magnetic film.

Married print Synonymous with a combined print in film.

Master An original recording, video or audio, e.g. master edit, master magnetic (film sound), master multitrack, master music etc.

Matching Equalizing sounds so that they appear similar, for example dialogue in discontinuous shooting.

Megahertz (MHz) One million cycles per second

Microsecond (ms) One millionth of a second.

Midi Musical Instrument Digital Interface, a technology that enables an electronic musical device to interact with others.

Midi Clock Midi code sent by a sequencer or drum machine to synchronize with other clock based Midi devices.

Midi Time code Time code produced by interfacing with SMPTE time code.

MIL Longitudinal measure equal to one thousandth of an inch.

Millisecond (ms) One thousandth of a second.

Mix In sound the combining of sounds. In vision, a dissolve.

Mixing consol See *Console*.

MMC Midi Machine Control.

MO Magneto-optical. A computer disk format.

Modular digital multitrack An eight-track miniature digital recorder capable of being linked to others.

Modulate   To vary from a mean position.

MOS, Mute   Shot silent, from German Mit Out Sprechen – without speech.

Moviola   Trade name for a film editing machine, an upright model usually favoured in the USA.

MPEG   Motion Picture Expert Group. A file compression for video and audio used in broadcasting.

MTC   Midi time code.

MTM   Midi tempo map.

MTS   Multichannel television sound. An American system of stereo audio compatible with the NTSC system. It consists of a pair of 15 K bandwidth channels, a separate audio programme channel (SAP) and a narrow bandwidth professional usage channel.

Multi-camera shooting   Using many cameras to cover a scene, all shooting at the same time; opposite of discontinuous shooting.

Multitrack recorder   An audio recorder with multiple recording tracks using one recording system.

NAB   National Association of Broadcasters. (An American organization.)

NC   Noise contour. A set of criteria used to assess the quietness of a room.

Nearfield monitors   Speakers used in close proximity to the sound engineer.

Neopilot   A synchronizing pulsed tape system without time code, compatible to pilot tone.

Nine Pin A   Sony protocol for interfacing editing machines. Now used on many audio and video machines for transport control.

NLE   Non-linear editing.

Noise reduction (NR)   A method of reducing the inherent background noise in a recording or transmission system. Notable types include Dolby SR; Dolby A; Dolby B; Dolby C.

Non-real-time   Faster or slower than real time, often referred to Digital signal processing.

Nowline   The position of the play head in a workstation.

NTSC   National Television Standards Committee. The initials NTSC usually refer to the colour television system primarily used in the USA and Japan, based on a 60 Hz mains system using 525 lines.

Octave . A musical interval spanning eight full notes. This is a 2:1 span of frequencies. Human hearing spans 10 octaves.

Off-line edit   The editing of material using low-cost equipment – for later assembly, on-line, with high-quality equipment.

Offset   The positioning, usually of a soundtrack, away from its normal synchronized position. Offset is measured in time and frames.

On-line video edit   Editing from information produced using an off-line picture workstation or video edit suite. This may be manually from notes, semi-automatic, or automatic from a computer disc.

On-the-fly   Working while the system is on the move. Choosing events points or edit points, or mixing without stopping.

OMFI   Open Media Framework Interchange. Manufacturer-introduced standard for interchange of audio, video and EDLs between different makes of workstations.

Oscillator   A device for producing pure-tone electric waves which are frequency-calibrated.

Overshoot To modulate a recording beyond its recommended level and thus introduce distortion.

Pad A fixed attenuator.

Parallel data In time code, time code in binary coded decimal form suitable for data output to an external unit; c.f. serial data.

Patch To make a temporary connection between fixed cable terminals at a (patch) bay by using a short length of cable called a patch cord.

PCMCIA Personal Computer Memory Card International Association; in effect a laptop manufacturers standard for memory storage.

Peak programme meter (IEC) An audio meter with a characteristic defined by the IEC.

Phase difference The time delay between two time-varying waves expressed as an angle.

Phase scope A form of oscilloscope having the ability to show phase relations between audio signals.

Picture lock The point at which no further adjustments to a picture are expected.

Ping ponging See *Bouncing*.

Pink noise Random noise with a frequency spectrum that appears to be equal at all frequencies to the human ear. Its random nature makes it representative of actual programme conditions. It can be generated by filtering white noise with a filter that has a slope of 3 dB/octave.

Polarity Positive or negative direction. The way of connecting a pair of signal conductors in phase or reverse phase.

Pre-roll The time allowed before the start of a programme or edit to allow the equipment, video or audio, to come up to speed and synchronize (up to 10 seconds).

Presence The natural background sound that a location possesses without dialogue or other spot sounds. Presence is used to fill 'holes' created during editing. Also called room time, ambience or fill.

Printing A term used in film for copying or dubbing off picture or sound, e.g. printing a soundtrack, copying it.

Print-through Transfer of one magnetic field from one layer of recording tape or film to the next on the reel. Analogue recordings suffer from pre-echo on tapes stored head out, post echo on tapes stored tail out. Archive material is particularly likely to suffer from print-through.

Pulsed Code Modulation (PCM) A system of recording digital audio.

Punching in/out The selection of record on a video or audio system, usually on-the-fly.

RAM Random Access Memory. Temporary storage on a computer, measured in megabytes or gigabytes of data.

Real time Actual elapsed time, also applied to workstation processing.

Record run Applied in time code to the practice of running code only when the recorder, video or audio, is in record mode. The code is therefore sequential on the tape despite the machine being switched on and off over a period of time. See *Free run*.

Reframing Reconstituting time code where the time code waveform is mistimed relative to the video waveform, making editing impossible. Video frame start and time code start are incorrectly related.

Re-recording, GB dubbing The process of mixing sound to the final mixed master.

Restored code   Replayed time code processed to improve its quality.

Reverberation   The persistence of sound in an enclosed space due to reflections from the enclosing surfaces.

RMS   Root mean square: a method of measuring level where the true power of the signal is calculated. Short spikes or peaks in the signal have only a small effect on long-term readings. The signal is squared, the average taken and the square root taken.

Room tone or presence   Background sound through a particular scene.

Rough cut   An early version of an edit.

Rubber numbers   A process of printing reference numbers on magnetic film and picture film to aid synchronization. (US) Ink numbers.

Run-up/run-out (USA)   The blank video tape or film used for threading at the beginning and end of a recording. No test signals or programme material should be recorded on it. (In Britain, called leader.)

Rushes   The first viewing of camera and sound material after it has been returned from film processing, or has been recorded.

Safety copy   A copy of a master to be used if the master is damaged.

Save   The action of commanding a computer to memorize data, in case of failure. Most workstations automatically save material.

Sampling   Measuring an analogue signal at regular intervals.

Scoring   The recording of music for a programme or motion picture after it has been edited.

Scrub   Moving sound back and forth over a point to find a particular cue or word.

SCSI   Small Computer System Interface. A standard for the interface between a computer's central processor and peripherals such as additional hard disc drives. Pronounced Scuzzy.

Separate sound recording   See *Double system recording*.

Sequencer   The brain of the Midi system.

Serial data   In time code, code with all 80 or 90 bits in correct sequence, i.e. data in a continuous stream from one source.

Server   The sender in a local area network.

Servo   An electronic circuit which, for example, controls the capstan speed of a video recorder. The servo may be adjusted by an external source for audio post-production.

Shoot to 'playback'   Playback of a soundtrack on location so that action may be mimed and synchronized to it.

Shuttle   Variable running of a video or tape recorder up to high speed from single speed.

Signal to noise ratio (S/N)   This can be considered in three ways:

1 The difference in dB between maximum permissible signal level and the noise that is present at the maximum signal.
2 The difference in dB between the maximum permissible signal level and the residual noise.
3 The difference in dB between a given signal level and the residual noise.

Silent drop in (punch in)   When actuating the record button, a silent drop in is achieved without any extraneous sound being recorded.

Slate   An announcement, visual or aural, that gives identifying information for a recording.

SMPTE   The Society of Motion Picture and Television Engineers.

Software   A computer term for the instructions loaded into a computer for directing its operation.

Spacing/leader   Blank film stock used to separate one sound physically from another in a soundtrack. In USA called leader.

S/PDIF   Sony Philips Digital Interface. A consumer version of AES/ EBU 3.

SPL   Sound pressure level. The measure of the loudness of an acoustic sound expressed in dB. 0 dB SPL is near the threshold of hearing. 120+ dB is near the threshold of pain.

Split edit, L cut   An edit where the sound and vision start or finish at different times.

Spot effects   Spot sounds that are not continuous and go with a specific action.

Spotting   Identifying picture points for sound effects or music cues.

Sprocket   The driving wheel or synchronizing wheel of a film apparatus provided with teeth to engage the perforations of the film.

Standard conversion   The converting of television pictures from one format to another, e.g. NTSC 525 lines 60 Hz to PAL 625 lines 50 Hz.

Stems   The constituent parts of a final mix.

Streamer   A sweeping diagonal line across a film or video picture to pre-empt a synchronous event.

Striping   Pre-recording a video or audio format with time code or a video signal, so allowing it to be used in post-production.

Tail out (Tails out USA)   Tape wound so that rewinding is necessary prior to playing – often recommended for audio tape storage.

TB   Terabyte – 1024 gigabytes.

TBC   Timebase corrector. An electrical device that stabilizes video pictures.

TC   Time code.

TDIF   Tascam Digital Interface Digital audio interface used by the manufacturer of a modular digital multitrack machine.

Telecine   Derived from the words television and cinema, a device for converting film to video.

Thin   Colloquial term applied to sound lacking in lower frequencies.

Three stripe (USA)   See *Triple track*.

Time compression   A technique where the play speed of a recording is increased above the recorded speed. A pitch change device lowers the resultant pitch back to normal, resulting in a shorter actual time but correct sound quality.

Transient response   The response of an amplifier or recording device to a very short fast-rising signal. A measurement of the quality of a recording system.

Trigger synchronization   Synchronization is produced by 'firing off' the slave device at the correct time; this may not be sufficiently accurate.

Trim   To add or subtract time from or 'trim' an edit.

Triple track   35 mm sprocketed magnetic film with three distinct soundtracks; dialogue, music and sound effects. Known as three stripe in USA.

Two-three pull down   The technique that successfully transfers film at 24 frames per second to video running at 29.97 frames per second.

U-matic   A standard three-quarter inch video cassette format, once used in industrial and broadcast applications.

Upcut (USA)   The loss of either end of an audio source, caused when an edit point is wrong, e.g. the loss of the first syllable of dialogue in a sentence.

User bits   Undefined bits in the 80-bit EBU/SMPTE time code word. Available for uses other than time information, such as identifying reel numbers.

Variable area soundtrack   A system of recording sound on film in which the track is divided into two areas, one clear, one black. Used in Dolby Stereo optical recording.

Variable density soundtrack   A system of recording sound on film, now obsolete, in which the track is of uniform density across its width but in which the density varies.

VCR (video cassette recorder)   Enclosed video tape, reducing damage and contamination.

Vectorscope   A visual display showing the colour portion of the video signal. It is used to adjust the colour saturation and the hue.

Vertical internal time code (VITC)   Time code for video – encoded into the video signal, it can be replayed when the tape is stationary since the video-head is rotating – an advantage in finding cues in audio post-production. However, VITC cannot be easily decoded at high speed since videotape recorders suffer from picture break-up in this mode.

Virtual mixing   A system for memorizing mix parameters.

Voice over GB   Commentary, explanatory, non-synchronous speech, super-imposed over sound effects and music. Also called commentary or narration.

VU meter   A volume units meter that measures the average volume of sound signals in decibels and is intended to indicate the perceived loudness of a signal.

Walla   A track containing no discrete speech, only a background burble.

WAV File   Windows Wave File, a sound recording system specific to a PC. A multimedia sound file that can be used within Windows with a variety of different sound cards and software.

Wavelength   The distance between the crests of a waveform.

White noise   Random noise with an even distribution of frequencies within the audio spectrum. This form of noise occurs naturally in transistors and resistors.

Wild track   A soundtrack which has been recorded separately from the picture, wild, without synchronizing.

Window dub   Time code placed as a 'window' on copies of source footage to help identify original shots in an off-line edit. GB: Burnt-in time code (BITC).

Wow and flutter   Variations in speed of a reproduced sound from the mean due to inaccuracies in the mechanics of the system, expressed as a percentage. Relates only to analogue recording.

Index

ⓕ Focal Press

http://www.focalpress.com

Join Focal Press On-line

As a member you will enjoy the following benefits:

- an email bulletin with **information on new books**
- a bi-monthly **Focal Press Newsletter**:
    - o   featuring a selection of new titles
    - o   keeps you informed of **special offers, discounts and freebies**
    - o   alerts you to **Focal Press news and events** such as author signings and seminars
- complete access to **free content** and reference material on the focalpress site, such as the focalXtra articles and commentary from our authors
- a **Sneak Preview** of selected titles (sample chapters) *before* they publish
- a chance to have your say on our **discussion boards** and **review books** for other focal readers

Focal Club Members are invited to give us feedback on our products and services. Email: worldmarketing@focalpress.com – we want to hear your views!

Membership is FREE. To join, visit our website and register. If you require any further information regarding the on-line club please contact:

> Emma Hales, Promotions Controller
> Email: emma.hales@repp.co.uk
> Fax: +44 (0)1865 315472
> Address: Focal Press, Linacre House,
> Jordan Hill, Oxford,
> UK, OX2 8DP

Catalogue

For information on all Focal Press titles, we will be happy to send you a free copy of the Focal Press catalogue:

USA	**Europe and rest of World**
Email: christine.degon@bhusa.com	Email: carol.burgess@repp.co.uk
	Tel: +44 (0)1865 314693

Potential authors

If you have an idea for a book, please get in touch:

USA	**Europe and rest of World**
Terri Jadick, Associate Editor	Christina Donaldson, Editorial Assistant
Email: terri.jadick@bhusa.com	Email: christina.donaldson@repp.co.uk
Tel: +1 781 904 2646	Tel: +44 (0)1865 314027
Fax: +1 781 904 2640	Fax: +44 (0)1865 315472